The Presentation of Original Work in Medicine and Biology

"Science begins only when the worker
has recorded his results and conclusions in
terms intelligible to at least one other
person qualified to dispute them."

B.M. Cooper, 1964

The Presentation of Original Work in Medicine and Biology

HUGH DUDLEY, Ch.M., F.R.C.S.(Ed.), F.R.A.C.S., F.R.C.S.(Eng.)

Professor of Surgery, University of London;

Director of the Academic Surgical Unit,

St Mary's Hospital, London

CHURCHILL LIVINGSTONE

EDINBURGH LONDON AND NEW YORK 1977

CHURCHILL LIVINGSTONE
Medical Division of Longman Group Limited

Distributed in the United States of America by
Longman Inc., 19 West 44th Street, New York,
N.Y. 10036 and by associated companies,
branches and representatives throughout
the world.

ISBN   0 443 01583 X

Printed in Great Britain by
Lowe & Brydone Printers Limited, Thetford, Norfolk

Acknowledgements

My interest in the presentation of scientific work was first aroused by the late Sir James Learmonth - ever a doughty champion of rigour in both speech and writing. Since then I have benefitted from criticism of my own efforts by a number of amateur and professional editors and particularly from Dr Ian Douglas Wilson, Dr I. Munro and Dr Stephen Lock. We disagree often and the faults and idiosyncracies are all my own. The challenge of running a department that was trying to communicate and the refreshing atmosphere of the Surgical Research Society of Australasia stimulated me to produce the first outline of this work. Mr. R.M. Kirk has provided me with many fresh ideas and suggestions. A number of my colleagues have permitted me to pirate examples of both good and bad writing. Professor K. Cox has pointed out many solecisms and if I have not always accepted his advice, that must be blamed on my bad judgement. My wife has not only kept me on the right lines, but also has typed and retyped my manuscript. Finally, two secretaries - Miss E. Duff and Miss S. Lind Jackson - will know how often I have departed from the advice I tender to others, but will I hope accept my thanks for their patience and endurance of many demands over the whole range of published work dealt with here.

Textual acknowledgements

For permission to quote material I am indebted to:

1. American Society for Metals and the Managing Editor, *Metals Progress*
2. Mr C. Battersby and the Surgical Research Society of Australasia
3. Professor L.H. Blumgart
4. Professor J. Calnan and Mr A. Barabas (Table IV)
5. Jonathan Cape Ltd (Table II)
6. Mr S. Desai
7. Dr B. Dixon and *New Scientist* (Table I)
8. Mr H.H.G. Eastcott
9. Medical Branch of the Library Association (Appendix 2)
10. Mr A. Nicolaides
11. The Editor, *World Medicine* (Table III)
12. The Editor, *Marine Week*
13. The Editor, *British Medical Journal*
14. The Editor, *The Scientist Speculates.* Wm. Heinemann Ltd
15. Professor P. Bell and The Surgical Research Society of Great Britain

And many others who wittingly or unwittingly figure in these pages.

Contents

Introduction

Of the making of books and articles, papers and communications there
is no end; meetings and the reports of their proceedings also multiply.
One consequence is a thriving secondary industry which explains, or
purports to explain, how best to communicate. Although there has been
much on this theme directed specifically towards medicine, I have not
been able to direct those of my staff who consult me towards a single
source of the information that they seem to need (if not to want) in
order to prepare themselves for delivery of a paper at a scientific meeting,
to equip themselves to write clearly, or to complete their theses for higher
degrees. This is not to say that we lack admirable books on various aspects
of the subject of communication; far from it, and in attempting a synthesis
I have drawn a great deal on others. What follows is the outcome of noting
down over the years in two busy academic surgical units what has proved
useful, what has gone wrong and what might be improved in the area of
communication. Gradually, a confused empirical dossier has grown up
which forms the skeleton of my present attempt. The flesh that can give
it form will only be provided by ranging widely over the literature. To
do so is not easy for those who are making a start in clinical science
because the reference matter of the subject is so scattered. I have attempted
to rectify this by a selected and doubtless biased reference list which
appears in Appendix 1.

One can argue that there are so many ways of doing things in the field
of communication that useful generalisation is difficult if not impossible.
It may be so, but many years of perpetrating solecisms as a writer and
correcting them as an amateur editor lead me to believe that there *is* a
canon of effective scientific speaking and writing about which generalisations *can* be made. True, this canon largely consists of what is left
behind when mistakes have been eliminated; it is less easy to give positive
advice than it is to correct error. Nevertheless, guidance can be more than
a sterile "do this, do that". Indeed, it is sad that English, one of the most
flexible languages there is, has been taught to most of us and interpreted
by many in a tyrannical way, particularly in relation to arbitrary rules of
usage. Diktats, and the wrath which tends to descend on the heads of
those who wittingly or unwittingly depart from convention, have been
used as the only mechanism of criticism. Attacks on a writer's or speaker's

grammatical or semantic quirks then get in the way of an informed analysis of what he is trying to say.

Concentration on criticism of method also has dangers at scientific meetings. The element of competition is always present, and rightly so; but this should be directed primarily towards the content of the paper rather than the technique of communication. To misuse the jargon currently in vogue: the *message* is more important than the medium. However, though visual aids, sonorous phrases and other embellishments will never cover up a basic lack of ideas and facts, fascinating work can be completely spoilt by a bad presentation which fails to catch the audience's attention or merely provokes irate criticism of style as distinct from content. The risk is the greater in a verbal performance where the listener has only one chance to make sense of what is said.

Few of us are naturally good communicators, but there are also few who cannot acquire, by dint of self-study and a modicum of diligence, sufficient ability to get by without failure even if also without undue acclaim. The objective is to leave the auditor or reader feeling that his interests and intellectual comfort have been studied. Sir Arthur Quiller-Couche[1] used to refer to the two courtesies - on the one hand by the speaker or writer to his audience and on the other hand by them to him. If one expects the latter one must offer the former.

Most of the previous writers on this topic have separated the spoken from the written word in the belief that they are two distinct areas which require widely different approaches. I have done so only in matters of detail because I believe there is an underlying unity in scientific communication which is independent of the vehicle used. However, some special features require individual treatment; these I have tried to specify. My unified approach may be a little clumsy, but I hope that the reader will emerge (if emerge he does) with a better conceptual grasp of his role as a communicator than he would otherwise have achieved.

The major criterion that has guided these notes is utility. Does it help? Does it work? Was it received well? These may be lowly objectives, but in trying to get ideas across the gap between persons I believe them to be fundamental. Language - what Pickering[2] has called the lost tool of learning - must largely have grown from utility. In the hands of many it has achieved a beauty and an artistic life all its own, but we should not be blind to the fact that, along with other tools of communication such as the much used and abused audio-visual aids, the prime purpose of what we do is to get a message from one to another in the most effective way. Efficiency and beauty may go hand in hand, but both may exist without the other. Our job is to ensure the first and hope for the second.

REFERENCES

1. Quiller-Couche, A. (1916) *On the Art of Writing.* London, Cambridge University Press.
2. Pickering, G. (1961) Language, the lost tool of learning in medicine and science. *Lancet,* **2,** 115-119.

1. Deciding to communicate

Why we communicate in science in the way we do has a long, complicated and confused history too diffuse to be considered here. Those who feel that they would like to know more of why, as practising investigators, they find themselves in the middle of a communications maelstrom should consult a recent publication by Meadows.[1] Suffice to say that there are both natural and unnatural pressures which drive us to make public what we do. The natural ones are: that scientific knowledge is not such until it is shared and subject to critical scrutiny; that such scrutiny is fun; and that the nature of the scientific creed imposes on us the responsibility of making the world rational to others. Less natural, but no less real forces are: those that lead to a feeling that a personal bibliography is an important tool for securing preferment; that consequently it is necessary to publish or be damned; and that competition between individuals or groups can be played out by rivalling colleagues in the number of presentations made or the time monopolised in discussion at a meeting. None of these is absolutely bad and all can contribute to the dynamics of scientific creativity; however, carried to the point of obsession, they may ultimately result in the fury of communication taking precedence over the conviction that there is something worthwhile to get across.

I do not wish to lay down arbitrary rules for deciding whether or not to communicate. Usually the decision should be a consensus among workers in a department, unit or team based on available channels and a feeling (not necessarily rationally based) that there *is* something to communicate. To underpin such a feeling Dixon's rules[2] are worth consulting (Table I). An additional and useful point to remember is that so-called negative results may be of value if they confute a hypothesis, provided this was a tenable one before the work was done. On occasion, innately absurd hypotheses are erected so that diligence can be displayed by their disproof. In biology this is not science - merely intellectual games. Nevertheless, because so much of medical practice is based on inherited empiricism (it works even if it is not the best or most economical or most rational way), it can be useful to negate a cherished belief. Length of stay in hospital, use of various established treatment regimens in ill-understood disorders are both good examples. Such activity in rejection of the conventional wisdom is in tune with the hypothesis-test sequence which will be discussed in more detail later.

Table I Dixon's rules for those contemplating publication of their work

---

Have I since my last publication felt any trace of anxiety or envy about the progress made by any of my colleagues? Have I hurried this paper in order to anticipate a publication by someone else? Is my work so inconclusive that I should wait some months before writing it up?

Does the paper describe anything new and useful? Or am I draining the last ounce of useless information out of a technique I developed for my Ph.D. work years ago?

Would I be stimulated if I read this paper in a journal under someone else's name? Supplementary: Would I pretend not to be? Would I pull it to pieces for the benefit of final year honours students?

Is the paper too long in relation to its factual content? Meditate on those who must read what you have written.

Have I written the paper to bolster my promotion prospects?

For authors of review articles: Is yet another review of this topic really necessary? Am I the right person to write it? Is it a truly critical review? Have I given undue prominence to my own work? Did I do it because it was a quick and easy way of extending my bibliography?

Have I even been guilty of the worst sin of all - staring glowingly at a pile of reprints of my old papers, of fondling them lovingly, making unnecessary lists, or even weighing them?

Does this paper, which will be published with my name at the top, represent my own work and nothing but my own work?

The choice of an audience for the spoken word is nearly always based upon speciality interest - only a few places and journals now survey the general field of biological work (The Scottish Society for Experimental Medicine is one and I suppose The Royal Society is another). However, the overlap of interest between societies does permit some latitude of choice. Unfortunately, it may also encourage a **practice** of multiple presentation which, for several reasons, is to be deplored. First, boredom for those in the audience who have heard the matter before; second, the impression that the speaker is on the make for travel expenses; third, if the paper is accepted in two places someone else has been unfairly rejected. Some societies make it a rule that the work submitted should be both original and presented only to them (see for example the Surgical Research Society abstracts in Appendix 3).

There will always be exceptions to these guidelines. A worker may be invited to submit because his material is so interesting that it needs to be heard. The audiences in different groups may not overlap. Nevertheless,

Dixon's rules (Table I) could be extended to include the following soul-searching before a paper is submitted to a meeting or journal.

1. Have I done it before?
2. If so, is the audience I now propose to enthrall similar or dissimilar?
3. Am I looking for a free trip?
4. Are we bolstering the Annual Report?

Choosing a journal

Beggars can't be choosers, and because research workers submit papers to journals which accept them graciously or refuse them usually equally so rather than the other way round, this section may appear unnecessary. However, to a minor degree constructive thinking about choice can heighten the chances of success. Little has been written about the subject, and as with examinations wild rumours do tend to circulate about editors, journal policy and the evil practices of referees.[3] Some of the individual tragedies of science are certainly histories of publication in the wrong place or of refusal by editors to publish at all. But these are probably rare in spite of rumours to the contrary that grow lush on the academic grapevine. Recently some, if not enough, editors have tended to be more forthcoming about their roles and policy. A rough classification can be made into general journals, those of a recognised discipline and those of a speciality or area interest.

General journals (such as *Lancet, British Medical Journal, New England Journal of Medicine*) take the whole field of medicine and its interface with biology as their stamping ground. In doing so their intent is to seek for a measure of immediacy and to support papers that look as though they might become of key importance in the development of knowledge or understanding. By contrast, they are rightly unwilling to waste precious space on things that are trivial or may prove sterile; and at the same time news, reviews and debate of the work of medicine or science as a whole must fill some pages. It is a daunting task - though ultimately their prestige depends upon the original papers they publish, and they must live in an uneasy balance with that type of material. Finally and inevitably, they tend - under pressure of space, deliberate editorial choice or because by doing so they can push out the boat of a new discipline or area - to develop themes (or at least to espouse them for a while).

Specialist journals in a discipline - for example surgery, internal medicine, endocrinology - trouble themselves less about day-to-day affairs; in a word, and perhaps loosely, they tend to be less journalistic. Their editors and editorial boards are usually volunteers and amateurs, continuity being provided by tradition and a publishing house in various proportions. Further, they may be the official or unofficial vehicles for the publication of the papers and news of a society, even if it is rare to find that this dominates their affairs. Nevertheless, if a paper accepted for the meeting of a society comes to be published automatically in the affiliated journal, pages of space may be blocked, so reducing the acceptance rate for those who are outside and prolonging the interval between acceptance and publication. The practice of publishing papers in full because they have been read at a meeting is also bad in its influence on the type of presentation at the meeting; it inevitably means that speculation and incompleteness are less likely, lest the paper "will not look good (i.e. orthodox) on publication". Also, some papers which subsequent to criticism at a meeting should not even see the light of day still get published. However, these observations only apply to a minority of specialist journals within disciplines. The majority provide a steady outlet for papers that are apposite. Their problem is to define that term and also to meet a certain critical standard as they try to bolster the status of both their discipline and their journal.

Finally and more recently, there has been a growth of journals which cover an area of interest. They are the offshoots of three things: the growth of scientific knowledge; the necessarily transient thematic activity of the general journals to which I have already referred; and the delays or rejections imposed by existing specialist journals. Some have already, or will in the future, move on to the status of specialist journals as the area becomes transformed into an established discipline (e.g. *Journal of Molecular Biology*). Others may continue under the wing of a society, though just now the economic possibilities of this look exceedingly bleak. Yet others will fold their tents and vanish. Journals which reflect an area interest, until they become established or that area moves down to the main field of development, are usually the easiest in which to get a paper accepted. By the same token they will be read by the fewest and the company kept may be the least inspiring.

What I have written is a very sketchy account of a complex subject that almost deserves a whole book to itself in the form of a Baedeker or better, Ruff's "Guide to the Turf". Nevertheless, some case law does emerge against which an author can examine his paper. Let me try an example, apologising as I do for taking the name of journals in vain and possibly quite genuinely misinterpreting their sometimes inscrutable editorial policy. Let us say that an author is writing about burns. If the paper is on "Tangential Excision of Burns: Methods and Results" it would be unsuitable for a general journal because it is a technical contribution directed solely at those who are doing the work. Therefore, a surgical journal, or even a quite specialist one dealing with plastic surgery, would be more appropriate. But burns are an area of interest rather than the province of an established discipline, so publication

in the *British Journal of Surgery* or *British Journal of Plastic Surgery*
might not be easy because of the pressure of other more discipline-bound
papers. That this has proved so can be recognised from the foundation of
a new journal - *Burns Including Thermal Injury* - and this would be the
first choice for our technical paper. By contrast let us say that the paper
is on the "Psychological Effects of Burns on the Body Image" and that
it carries a general message for anyone who has to look after patients who
have been mutilated, as well as illustrating the existence of a previously
unrecognised problem. Then it would be appropriate to consider submission
to a general journal - say the *Lancet.* In between these two extremes a
paper on "Stress Ulceration in Burns" which illustrates a general principle
within a discipline (in this case the physiological response to injury) might
go to the specialist journal within that discipline - say the *British Journal
of Surgery.*

In general, the two main points that must be answered are: What was
the objective and what has been achieved? The two are not necessarily the
same by any means. They can determine which audience is likely to be
receptive and this in turn will enable the emphasis of the paper - and
particularly its discussion - to be correctly stated for that audience. Some
may say that this is carrying the journalistic approach too far into the
production of scientific work. I do not agree. What has been done will
ultimately stand or fall in its own right; to present what is believed to be
useful or new in a manner which is appropriate to the overall pattern of
the chosen journal is therefore nothing but common sense.

Choosing a forum for verbal presentation

One can either enter a magic circle or one cannot. Societies and clubs
promote papers through their members and by a more or less rigorous
system of judgement on abstracts. The appropriate place will be a
combination of that to which one has access either directly or through a
colleague (who in that old-fashioned but gracious term can "introduce"
the stranger) and the relevance of the communication to the aims of the
society. The latter are often difficult to judge, but similar classificatory
rules apply as for journals. There are very few general places in which
short scientific papers can be presented - the field is now too large - and
even interdisciplinary societies are finding it increasingly hard to keep their
members together as areas of special interest and understanding become
the preserve of one ingroup. Circumstances, interest of the team with which
the author is associated, availability and luck seem to be the factors which
will get a paper into a scientific programme. In addition, the quality of the
abstract may be all important. Writing an abstract, which is to a degree an

art rather than a science (that is, I cannot easily dissect its components), will be considered later.

I have no comments to make on large, international meetings for which it seems possible to get nearly all papers accepted without difficulty. The implication for me is that these meetings are rarely worthwhile except as a method of travel and the opportunity to meet individuals outside the lecture theatres. Perhaps this is too harsh a judgement, but in writing about scientific communication at meetings I am limiting myself to those which take place at a domestic level and which, when one is optimistic, one hopes will provide the model for international communication in the future.

REFERENCES

1. Meadows, A. J. (1974) *Communication in Science*. London, Butterworth.
2. Dixon, B. (1967) Mirror, mirror on the laboratory wall. *New Scientist*, **36**, 712-713.
3. Intercepted Letter (1973) What do editors want? *Lancet*, **2**, 435.

2. Structure of scientific thought in relation to spoken and written communication

A persuasive case has been made for saying that most scientific communication is to a degree misleading if not, as has also been claimed, fraudulent.[1] The style of both spoken and written papers is certainly based on a retrospective ordering of ideas and a shuffling of the way things were done, which then produces the conventional sequence of introduction (including a review of previous work), methods, results, discussion. Often, of course, and particularly so when novel ideas are being introduced, the matter has not been so simple. Vague concepts may lead to tentative tests, often ill-controlled or loosely designed. If these are fruitful then a more formally structured experiment is set up; it is this that then makes the body of the scientific communication. Development of ideas in time is often lost to both author and reader and with it may go something of the excitement of scientific discovery, an excitement that on occasion can be enjoyably communicated to others.

This is not to say that scientific investigation cannot be close to the structure of a scientific communication or vice versa. If, for introduction we substitute *hypothesis* (either original or derived from the synthesis of the work and ideas of others), for methods and results *test,* for discussion the *logical results* that may be deduced from the experiments performed plus any *inductive generalisations* that seem appropriate, then we have a sequence of events which in part follows a currently popular view of scientific process - the hypothetico-deductive approach.[2] The meaning of the word *induction* as used here is assumed, because a more detailed consideration would take us into a debate which, though philosophically rich, is of tangential relevance only; those interested may care to consult two recent accounts.[3,4] If, in addition, the discussion contains some hint about how the original hypothesis should now be modified or rejected we have closed the circle of scientific inference (Fig.1). The scientific paper, read or written, is then a manifestation of the ordered and conceptual thought processes that have ultimately taken place, even if it does not necessarily follow the chronology and the stumbling efforts that have gone into the realisation of the concepts. As such it is certainly not fraudulent in ideas.

Two practical results emerge from this discussion. First, the conceptual framework of Figure 1 should be used to test the structure and context of what is written or said. Second, although much of the sleep-walking may consequently be lost, some of it, rightly communicated, may be interesting, instructive and humbling for the reader or listener. Though inessential to the ultimate flow of ideas it is highly relevant to an understanding of the blocks and dams that may impede in real life. Finally, it is humbling to look back in that the vagueness of what was originally conceived may then be exposed.

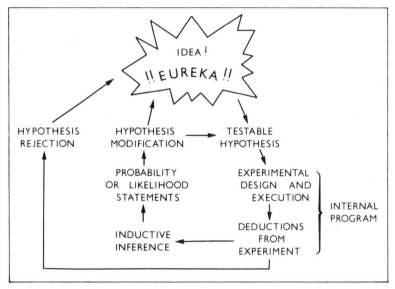

Fig. 1 The circle of scientific inference.
This is a stylised view with apologies to Sir Karl Popper.

Not all scientific communication results from either conscious or intuitive recognition of the hypothesis-test principle. New information may be the only goal, particularly in areas where knowledge is limited or where the advent of new techniques permits measurement to become precise. Some have been sceptical of such abstract empiricism (see Wright-Mills[5]) and it is indeed probably true that the gathering in of data without a precise guiding hypothesis (which can be called a Baconian approach) is likely to be less productive of new ideas. The same might also be true for a lot of confirmatory work. Nevertheless, apart from mathematical thought, we do not commonly work in a factual vacuum and data-gathering undoubtedly has its use.

If a paper is to be written on facts alone then it is important to say so. By being honest one can be more certain of gaining interest. Further, it is often possible to shorten the labour of writing and thus to produce a briefer paper. It is no use straining at the gnat of hypothesis if it is unimportant.

However, it is obviously wise to choose a productive area from which to gather data (in this, one *is* entertaining a hypothesis about the data, but in a slightly different sense from that usually meant). Also, getting information should be fun.

The impact of scientific inference on the form of communication

Examination of the work done in terms of the above should result in a form beginning to emerge for a written or oral presentation. The objective either to test a hypothesis or to gather new data - must be defined and stated. If the first, it is necessary to give a brief background to the current state of the hypothesis. Chronological features may be included both from the literature and, if this is not the author's first report, from personal experience. Selection is almost inevitable, and the possibility that an omission will offend against the history of science in general or against an individual in particular must be accepted. There is a courtesy in including the ideas of both living and dead and nothing to be lost by doing so; the value judgement lies in deciding when it was *ideas* that were contributed. The introduction that traces a detatched or exhaustive chronology of every contribution to the subject is rarely read, but some attempt must be made to review the events that have led up to both the current state of the play and the nature of the author's personal involvement.

If the second - the gathering of new data - the only thing to be firmly stated is *why* this is needed. It may be because two or more sets already obtained are in conflict; or because there is nothing in the field at all (rare these days); or because, as already mentioned, new techniques refine the precision of measurement. In the last case, care should be taken to try and assess the worth of what has been done before rushing into print; new figures obtained with a fine piece of apparatus that just happened to be lying around are valueless if they do not increase the precision of results already available. Of course, that work should not have been done, but the fact that it has been is no excuse for presenting it.

Defining the objectives writes most of the introduction. To summarise, it is worthwhile quoting the Editor of *Nature* who has recently said:

> "If more authors ... would attempt to make the first paragraph into a crystal clear description of what the paper is about rather than what other people's papers have been about, *Nature* would be an easier journal to read."[6]

Statements of intent and those that satisfy expectations are also valuable. A sentence or even a short paragraph at the end of the introduction which says what the work has or has not achieved as well as what it set out to achieve may catch the reader's attention and invite him to test for himself by further reading whether or not he agrees with the contentions expressed. Two examples are:

"We report here results that support/confute the above hypothesis."

"We have now obtained data which permit more accurate measurement of ... and thus more precise care for the patient."

All that has been written about the introduction could apply equally well to a summary. Indeed, if one reads the summaries of papers in (say) the *Lancet* one can often get the sense of the paper without reading the introduction. Were it not for the chronology and development of ideas which a good introduction can and should contain, it might well be abandoned or at least condensed into a summary. It is certainly worthwhile seeing how much of the summary will do for the introduction and vice versa. The converse of this argument is that a good paper is its own summary and that the latter is unnecessary. This is my personal view, but currently it does not seem to be shared by many editors.

Methods are usually the easiest part of the paper, though for a brief oral scientific presentation care has to be taken to include only what is absolutely necessary. In either case it is highly desirable to remember that

(a) it must be possible for those in the same field to follow the steps exactly to reproduce the results; and

(b) less initiated people should be able to understand the *principles* of what has been done.

Particularly in presenting a short scientific communication by word of mouth these objectives conflict and precise rules cannot be specified. One piece of advice is to avoid drawing up descriptions of techniques in code words which, though perhaps known to the initiated, are unintelligible to the majority and thus only deepen mystification. It should also be said that the house rules of a journal or the traditions of a society must be studied with care. Finally, it is easy for a referee or a selection committee to pick holes in the section on methods with consequent irritating delays or even rejection.

The basic rules for methods in both written and spoken word are:

1. How much detail? Usually only critical steps which illustrate principles should be described.

2. How much justification? Validation on test material, assessment of reproducibility and of the range of error may all be important to the effectiveness of the method for the purpose to which it is put. It is a fair criticism that these matters are less often dealt with in medical scientific work than they should be.

3. How much reference and in what form? If a technique has been handed
down from worker to worker and much modified in the process it is
good scientific history, but wasteful of precious space or time, to
detail every step in its development. First and last are usually sufficient
unless the intermediate changes have been vital. To describe a technique
by name may also be a monument to a distinguished predecessor but
should not be done without the appropriate explanatory qualification.

All that has been written above about methods, their justification and
description applies equally, in my view, to statistical methods. Many
biological scientists, particularly those associated with medicine, are just
beginning to acquire the modes of behaviour that have long since characterised
their colleagues in agriculture, psychology and sociology. To describe the
statistical method used and to say why it has been chosen educates others
and displays probity. Apart from this, there is nearly always someone in
the audience or among the referees or readers who has made a special study
of statistics, often it must be admitted with the aim of trapping the naïve.
Therefore, it is a matter of survival to have a grasp of the statistical methods
used, even though advice may be taken from an expert on their choice.
Acquisition of statistical wisdom is far from easy, but I have included a
bibliography which may help (Appendix 4).

There is an increasing tendency to use statistical techniques as a method
of confusion and therefore to promote one-up-manship - part of a more
general phenomenon of creating a discipline with its own code (jargon)
which has to be broken by the would-be-initiate before he can be admitted
to the inner circle. One recent comment that I enjoyed - on a paper from
my own department- was:

"The methods and statistical analysis were most elegant but no one
could get within a mile of the raw data."

Before statistics became the chief tool of obfuscation in my own field
it was the language of transplantation and before this that of gastric
physiologists. Perhaps this is inevitable at the growing edge of a scientific
pursuit where definitions have to be created to cover new ideas or facts,
where initially meanings are known only to a few and the semantics are
fuzzy at the edges. However, a great deal of confusion could be avoided if
those in the know would translate for those who are not. In this context,
and more generally, it is time well spent to give the definition that it is
proposed to use either in a glossary to the methods section or, in a longer
work, in a section which precedes the whole text. In spoken communication
an introductory slide can serve the same purpose. For example:

1. In a written presentation "the symbols and abbreviations used in this
thesis follow the pattern laid down in *Biochem. Jour.* (1973) **131** 1-20,

Policy of the Journal and Instructions to Authors."
Additional abbreviations are as follows.
   HSV: highly selective vagotomy, a procedure applied to the body
       of the stomach only.
   TV: total division of the intra-abdominal vagi below the diaphragm.
2. In a spoken communication these two definitions, HSV and TV,
   would be put on a slide before the results are introduced.

Results

In both spoken and written communication results are a central section
and one which requires the most care. A middle course has to be steered
between, on the one hand, the presentation of raw and indigestible data
which cause reflex rejection and, on the other hand, the temptation to
synthesise to such a degree that the original information is lost. One guide
is to consider if the data can be reasonably presented in tabular rather
than textual form. Another is to ask: "Is this item essential to my interpretation
and conclusions or am I putting it in just because it is there?" A third
solution to the data problem, and one not used often enough in medical
literature, is to report that the raw information is filed in a library and can
be got at by anyone interested in working it over (see for example the
practice of the *Biochemical Journal*[7]).

Whichever tactic is adopted, it is desirable to try as far as possible to
permit the recipient of the communication to rework the data if he so
wishes. The results will often surprise and occasionally enlighten.

Discussion and conclusions

These will by now almost have written themselves, particularly if the
introduction is couched in the way I have suggested. However, a word of
warning is necessary. The model used for scientific inference - hypothesis,
test, deductive inference, generalisation, new hypothesis - is circular.
Therefore, the circle may be gone round more than once in a discussion.
By this I mean that if the results lead to certain generalisations then further
results should be obtained which would lead to another set of inferences,
and so on. This process of accumulating untested hypotheses is fun; it
has a place in a speculative review; but it should be used with caution in
an ordinary scientific communication. Once round the hypothesis -test
circuit is usually as much as the experiment and the data will stand.

If the results lead to self-evident conclusions it is only necessary to sum
up and to try and place the inferences in the context of the original hypothesis
or problem that has been posed.

Although I have urged caution in going round the hypothesis-test circuit it is desirable from time to time to be highly speculative. A curse of scientific orthodoxy is its caution. Both in written and spoken communication (particularly the latter) there is sometimes a case for chancing one's arm. There is nothing more disconcerting than to have delivered or to have had published what is thought to be a good and exciting paper only to experience a lack of response. There is always a tendency to blame the recipients ("I'm ahead of my time; how can you expect that lot to understand; I shouldn't have published in the -----: no one reads it"). The fault lies not in the audience's star; it is more often in oneself for having failed to be adequately stimulating.

REFERENCES

1. Medawar, P. (1964) Is the scientific paper a fraud? in *Experiment*. London, BBC publications. (This paper has been published in several forms and places - an example of the multiprint syndrome - and its general conclusions are also touched upon in the same author's *Induction and Intuition in Scientific Thought* (1969) London, Methuen.)
2. Popper, K.A. (1958) *Logic of Scientific Discovery*. London, Hutchinson.
3. Mackie, J.L. (1974) *The Cement of the Universe*. Oxford, Oxford University Press.
4. Swinborne, R. (1974) *The Justification of Induction*. Oxford, Oxford University Press.
5. Wright-Mills, C. (1959) *The Sociological Imagination*. New York, Oxford University Press.
6. Nature (1975) Editorial. Its your journal. **252**, 337.
7. Biochemical Journal (1973) Policy of the journal and instructions to authors. **131**, 1-20.

3. Matters of detail

Physical aspects of the preparation of manuscripts

Everyone has his own idiosyncracies about the preparation of a draft.
Some write, some type, some dictate (but see Dictating Machine Style,
p.48). Some work from notes, some straight out of the head. Given that
the framework of what is to be said and why it is to be said has been
determined and, if given further, that a structure of communication such
as that already outlined has been accepted, then it matters not at all how
a draft is produced. What is of concern is how that draft is transformed
into a final manuscript. Few of us have enough secretaries that we can
afford the luxury of repeated drafts, and in any case lavish help is something of an invitation to intellectual idleness. Therefore, the objective
should be a rough draft produced by the author, followed by one draft
typescript which leads directly to the final copy. For this ideal to be
achieved, which regrettably it rarely is, I have found certain matters must
receive attention.

1. Reasonable legibility in a hand-written draft, particularly if the typist
   is unfamiliar with the script.
2. A correct order of manuscript for the typist to work from. By this is
   meant that inserts and transpositions should not be signalled by lines,
   arrows or other calligraphic devices. Rather a manuscript should be
   attacked with a scissors so as to get things in order or to open up gaps
   to allow a new piece to be inserted. When this is done two further
   small matters deserve attention.
   (a) The cut up pieces should be stapled or fixed with transparent
       adhesive tape to backing sheets of uniform size. Otherwise it is
       difficult to handle the manuscript and small bits of text will
       almost always fall out.
   (b) Renumbering of the pages is essential and bold pencil numbers
       should accompany every manuscript.
       If these rules are observed it is unnecessary and undesirable to
       staple the sheets of a manuscript together. A typist will usually
       prefer to turn the pages on her desk from one pile to another; a
       stapled collection makes this impossible.
3. The author's draft should be as complete as possible. None of us is
   immune from the temptation to leave gaps for the references in the

17

text, the results in tables, the legends to figures; our desire to get on and see a typewritten draft may even lead us to omit parts of the argument on the assumption that they can be filled in as corrections later. All too often this produces a situation in which the structures of the paper becomes untidy and, as important, when the additions are made they alter the work so much that a further draft is needed. The same rule applies to grammar and style. So far as possible, aim for the final form rather than think "I can polish this in the next draft".

Finally, always have figures or illustrations made at least in rough form before a draft is typed. It is remarkable how their interaction with the text may require alterations to one, other or both.

4. Do not have references typed in a draft. Transcription of references nearly always leads to error. Furthermore, the physical labour is considerable and the reference list is almost certain to change before the paper is in its final form. If references are kept on cards (see p.26) it is easy to make up a pack of these to be typed as the last thing to be done before a paper is despatched (see also below, the final manuscript).

5. However legible the script, and even if it is impeccably arranged and paginated, it is useful for a secretary accustomed to a tape recorder to have an audio version to work from. To hear the words as the writer thinks they should sound may help her to punctuate consistently, if, as is common, that is left to her. Also, uncertainties about words can sometimes be clarified in this way, which is particularly helpful when the author is unobtainable. Moreover, to read what has been written into a dictating machine may help to ensure that the work *is* readable. Should it emerge that this is not so and changes are made on the recorded version, then the text must be made consistent with it or annotated to ensure that the spoken, rather than the written, word is followed.

6. Extra copies are wasteful of time and money; do not ask for them unless circulation to co-authors or critics is absolutely necessary. However, a secretary will certainly want to keep one copy so that when the draft manuscript is lost she can triumphantly come to the rescue.

7. Correct a draft on the top copy even if a carbon is available. It is easier for a secretary to work from the more legible manuscript. The paper of the top copy will usually take the pencil used for corrections more clearly than will a flimsy. By contrast, do not correct the top copy of a final draft - a good and authoritarian secretary should not even let the author have it lest he makes a mess of it. Use a soft pencil (B,2B, or 4B) for corrections, particularly of the final manuscript, and thus be able to erase on the carbon, so preserving a fair copy for

comparisons with a proof. Observe point 2 on a corrected draft.
8. Refrain from mutilation of a final manuscript by heavily applied paper clip in the upper left hand corner. It looks horrible.
9. Number photographic prints with a soft pencil on the back. State your name and indicate the top by an arrow even if this is patently obvious. Store the prints in the semitransparent envelopes in which they usually come from the audio-visual department. *Always* write the author's name on this envelope with a felt pen or some device which will ensure permanent identification. Anyone who has been a referee will realise that the jumble of papers received from an editorial secretary may result in illustrations of this kind being lost unless they are positively identified.
10. *Before* any drafting is done, *always* read the house rules of the journal to which it is proposed first to submit the paper. This may save much tedious correction at a later date.
11. Try if possible to store the reference copy of the paper in a file so that
    (a) it can be found,
    (b) it is preserved in good condition,
    (c) correspondence which relates to it can be referred to easily.
12. Try to present to anyone from whom advice is required a draft that is reasonably tidy. A dog's breakfast of tatty bits of paper is an invitation to take an adversary view or to skimp the critical task merely because it involves too much labour and frustration.
13. Do not fold manuscripts for transmission by post. Not only does this spoil their appearance, but it also makes them more difficult to read.

Authorship

A paper is to be presented or submitted for publication; who subscribes their name is then a vexing problem. However, it is an important one and should be debated, because it can cause a good deal of anxiety, particularly amongst junior staff. I have known it to rend apart a previously closely knit department. The following is a description of possibilities with some of the pros and cons of the various conventions.
1. Only the individual(s) primarily responsible for doing the work. Back-up support from the head of the department, except in the acknowledgements, is excluded (see Acknowledgements, p.21). As a subclassification of this either
    (a) authors in alphabetical order, or
    (b) he or she who is regarded as the prime mover in the work is named as the first author.

On the whole this general method - particularly its second variant - seems the fairest. There are two disadvantages. First, the effort required to sort out whether there has been a conceptual contribution which warrants the inclusion of the names of someone who did not do much of the real labour; in decisions of this kind departmental heads and others who direct the affairs of a research group are prone to feel that they have been left out. Second, the failure to include the name of a group leader who is well known may wrongly condemn the work to undeserved obscurity.

2. The name of the group leader (often the departmental director in clinical medicine) is always included. As a subset of this, anyone who is or was remotely connected with the project also has his name tacked on. Most group leaders seem at least to have the grace to put their names last when this convention is followed, but there they may stand out as starkly as elsewhere.

The advantage of this approach is that it does provide the (specious) publicity which would otherwise be lacking. Also, no one, least of all a paranoid professor, can cavil about exclusion. The disadvantage is that work done by others may unfairly be attributed formally or informally to the best known name (see Eponymous Reporting, p.22). A good deal of suppressed animosity comes to be generated against a senior member who insists on formal recognition in this way. In due course he may attract less effective workers and the base from which he operates may then crumble away.

The inclusion in the author list of everyone on a project as a means of expanding the bibliography of each is perhaps less common in Britain than elsewhere. It fools few people and is self-defeating. At an interview the first question likely to be asked is "What was your role in the paper in which you were co-author with the other nine?" Further, by studying the structure of authorship in relation to the subject matter it is usually possible to make inferences about role. Finally, a round dozen of other contributors, often with their titles and degrees, wastes space in journals. The largest number of authors I have encountered on a clinical paper is twelve[1] but I have not made an exhaustive study of the subject. If high energy physics is a model, then as the size of research groups in biology increases, so shall we see an exponential rise in authors.[2] Numbers in the fifties are already on record. Anonymity may be the only solution but I leave that debate to others.

We may conclude that in the choice of authors, as with many things, a degree of moderation is called for. Often, if the matter is frankly discussed, the problem disappears. Personally, I deplore the use of a leader's name on every paper but this *is* a personal view.

Acknowledgements

There is a tendency, particularly but not exclusively in American literature, to include every grant-giving agency in the list of acknowledgements, however irrelevant to the paper they may seem to be. It is true that we all rely on our grants for basic support (i.e. the purchase of stationery, technicians, alcohol and free travel) and therefore each and every research project is underpinned by *all* the research monies that accrue. Therefore, in theory, we should acknowledge everyone. A not unusual device which does stop the editors of journals from having hysterics is the use of the statistical convention which ascribes variation to real effects and remainder so saying "the laboratory is supported *in part* by etc., etc." Sometimes a strident note is detected - from where is the other part coming? One has difficulty also in escaping from the feeling that some of these acknowledgements are made to impress others rather than because there is a real obligation towards the grant-givers. Long code names and the invocation of agencies which carry weight in scientific council are likely to be cited with some pontification. Here is an example - what on earth does it mean?

> "This laboratory is supported in part by U.S. Public Health Service Grant ... from the National Heart and Lung Institute and by U.S. Public Health Service Grant ... from the general clinical research center's branch of the Division of Research Resources."

It is unlikely that lack of support from the same people will be recorded, although this might give us about the only check on the success that attends the guessing game always involved in the distribution of research grants. Perhaps contractually organised research will introduce more direct statements. If we test a drug why not say so? If we evaluate a hypothesis why not admit it? It is coy to the *nth* degree to say that research was supported *in part*, as though we want all the credit but none of the responsibility. Better to say, "we were asked (or we negotiated) to test this or that and here are the results."

Personal acknowledgements

These often make me, and I hope others, squirm. As John Vaisey has said, "a grotesque and fulsome list of indebtedness" is anathema to most in Britain, if not elsewhere. I like to be acknowledged if, as head of a department, I have helped, but the paper itself should fulfil that longing. Direct specific help should of course be recognised; non-specific support can be left out.

Eponymous reporting

By this I mean the ascription of a disease, a technique or some other event to one or more individuals. Often to begin with this has been both a nice gesture and a convenient shorthand; but it has disadvantages.

(a) The automatic inclusion of the best known name in the institution/ department/division (see earlier, Authorship) may generate a version of the Matthew effect (to them that hath shall be given - introduced to science by Robert Merton) so that however little - or much - the head man has had to do with the work, it is all ascribed to him. He will find it difficult to resist, and before he knows what is happening (perhaps) he is dining out, lecturing away and gaining academic mileage from the situation - the last performance recently described as a variant on Thorsten Veblen's conspicuous consumption - "conspicuous travel."

(b) The convenience of reference to a patient's disease, or to the subject, by the name of the initial author can have the same result. To say X *et al.'s* case is less easy on the eye and ear than to say X's case. So those aspiring to the Matthew effect should make sure they appear at the beginning of the list of authors.

(c) The long-established custom in medicine of referring to patients as the property of clinicians may bias the contribution or commentary thereon. Then if X,Y or Z are reporting and X is the clinician it is X's patient even if he made the crassest error in management from which he was saved only by Y or Z. Here is a recent example - without the error, I hasten to add.[3] There are eight textual references to another paper which had four authors: all but one of these refers exclusively to a clinician (X) as follows.

1. X or his colleagues.
2. The X report.
3. X's experience.
4. X's report (twice).
5. X reported.
6. X's patient.
7. X's discovery.

Maybe this example is so gross as to be a caricature, but the habit is degenerate and should be overcome. Getting rid of names altogether from the text is often easy but may cost a few words if the journal does not use superscribed numbers for references. Here are some possibilities

The patient (X *et al.*, 1974)
The patient previously described[29]

Others are easily invented to deal with situations that involve ideas and

these variants are certainly preferable to littering the text with the
names of individuals. Additionally, to refer to "Freedman *et al.*'s
work (Freedman *et al.*, 1944)" or to use "Freedman *et al.*'s work
(1944)" is inelegant and may be distracting from the main theme.
To get names into brackets or to use only superscripts requires
conscious effort, but undoubtedly makes a text easier to read.

All this is not to be taken as a retreat from the recognition of individual
talents and contributions where this is unequivocally appropriate. The same
issue of the *Lancet* from which the last example was drawn contained a
paper on a new and valuable technique.[4] It was properly introduced as follows:

> "G.W. Hounsfield working at the Central Research Laboratories of
> E.M.I. Limited in England has developed a revolutionary method of
> X-ray diagnosis."

Historical references

The acknowledgement of predecessors in ideas is gracious but may be
incomplete because of ignorance or dislike. Efforts should be made to
reduce the one and allow for the other. However, the literature has
become so vast that some omission is almost inevitable. Indeed, it may well
be asked if this matters. As temporal distance increases we must all (unless
we are Newtonian or Hunterian in stature) find our contribution much
diluted in the mainstream of ideas. The best we can ultimately hope for
is that we have contributed a drop to the flood. Indeed, as I write I am
conscious that I have read someone say exactly what I am now saying
somewhere else, at some time, but I cannot remember and thus cannot
acknowledge. Better to be guilty of the sin of omission than to adopt the
corrupt tendency of making the references for a paper by the appropriation
of whole chunks of reference material for quotation from other papers
without this being checked or read in the original. Scientists could learn
a lot from historians about the use of primary sources in this context. Chagall,
in a paper which, if bitter, should yet be read by all those interested in
communication in the biological sciences,[5] has drawn attention to this
debased tendency and its outcome in terms of a recognition of real contributions.

Notwithstanding the dubious worth of long-range historical citation and
its eventual defeat by the infinite capacity for the literature to grow, it is
conventional and conventionally virtuous not to miss out the living or the
recently dead.

Self-quotation

Auto-reference requires great care if it is to avoid self inflation and, unwittingly or not, give an impression that the writer has been more important to the development of an idea than was in fact so.

"We (or I) have previously shown that ..." followed by a reference is a perfectly proper way of starting a paper or a verbal communication which takes up where a previous one left off, though I must confess that I find "In a previous communication to the Society we showed that ..." a trifle pretentious as a way of starting to speak. By contrast to these harmless practices, consider the following written by Z (authors' names have been suppressed):

> "Evidence from this unit and elsewhere has shown increases in the activity of glucocorticoids, growth hormone and catecholamines (X *et al.,* 1966; Y, 1972; Z *et al.,* 1974) in the post-operative period."

On more careful examination only the last reference is to work by the author from his unit and indeed is listed in the bibliography as "in preparation". Therefore, the text places an inappropriate emphasis on the way in which the evidence has accumulated. Unless the reader looks up the reference list or knows that X and his colleagues were not associated with the present writer, he would be inclined to think that the last was first in the field. This is mild scientific dishonesty - not important maybe, but still to be avoided.

Unpublished work

The habit of referring to other papers written but yet unpublished, in preparation, or to be published elsewhere has the same flavour and seems to be on the increase. Even such august journals as the *Lancet* now allow it in the writing of leaders. Most of us are guilty of this practice from time to time, but it is proper to question its motives in ourselves as well as in others. In addition to the obvious situation that the results have not been completed, collated or organised, or that the paper still exists in the mind's eye of the department head, there is often in statements about future publications the implication that the writers have achieved a great deal and are about to burst forth in a number of places. A further motive is the separate publications that can be achieved by working over the data or various aspects of the study in different ways.

Even if motives are impeccable there is something very annoying for the reader in the statement that "these results will be published elsewhere". Where? How are we to find out? The same irritation and uncertainty attaches to a text reference which, when one looks it up, proves, as in the example given above, to be "in preparation". How long will this preparation last? I do not know of any figures which relate the number of times this statement is translated into a publication, but from my own experience I would hazard that less than a third ever see the light of day. The one exception to the deprecation of this practice is when the paper has in fact been accepted and is indeed *in press.* This gives the would-be reader a real guide to where it might be found. If he reads the first paper long after it was published he can follow a reference to "*Lancet*, in press" by looking up the yearly index of the journal for the following two years.

Finally, in relation to self-quotation, if one becomes involved in debate (often a euphemism for argument) it is well to have authority from others as well as oneself to back up statements or claims. If one fails to do this, this type of riposte is invited.

> "Dr Engel is entitled to his opinion but I doubt if he is engaging in an acceptable dialectic when he quotes himself three times as an authority substantiating his own argument."[6]

It did not matter that Watson was not himself on sound ground. His point was well made that it could not fail to demolish Engel.

The rise of "personal communication" comes under the same rules as "unpublished" and should be avoided as much as possible. The *Biochemical Journal* forbids it except in the text. It smacks too much of an "ingroup" making little private communications to each other which are more privileged than public utterance. The matter of science is public.

Reference conventions

References are irritating but important and some of the pitfalls associated with their use have already been dealt with. Methods by which articles or abstracts are referenced are governed by the house rules of the journal or society to which the work is to be submitted. In passing, we must note that the Harvard system (see below) or one of its modifications in which the authors' names are followed by date of publication (e.g. Smith and Jones, 1965) is easier for the writer than is a superscript (e.g. Smith and Jones[6] said, or it has been said[6]). For editors the superscript is more economical and disturbs the flow less, but it can cause chaos if there is a late insertion. However, the way references are kept and how they are inserted into a draft manuscript is under personal control. The Harvard system is the most complete and flexible, though for these reasons it is also the most clumsy and makes the greatest demands on the patience of the author and his secretary. An excellent description of the Harvard classification and its use is given in the publication of the Medical Branch of the Library Association who have kindly allowed me to include this as an appendix (Appendix 2).

To keep references for future use or to assemble them for publication, a card system is recommended. Though many journals will abbreviate references by omitting titles, by not quoting the closing as well as the opening page numbers, or even by condemning to the anonymity of "*et al.*" all those other than the first author who have laboured so hard to get their names in print, one should, in an index system, *always* copy the full reference on to a card so as to save another look for a missing bit of information. Edge-punched cards can conveniently be used to record the initial letter of the first author's name, so making it possible to arrange a batch of references into alphabetical order for typing up in a reference list. Alternatively or additionally, a temporary pencilled number can be used on the card when the journal's system is a numbered superscript.

Many of us have had grand designs for keeping all our references in this form using other parts of the card for subclassification of areas of interest. It has certainly been my personal experience that such elaborate schemes usually collapse over the years unless run by an obsessive personality, but they can be useful, particularly in that additional information may be entered on the body of the card. For a limited task such as a book or thesis which nevertheless requires a lot of references that have to be shuffled around a great deal, an edge-punched system is strongly advised. Of course, computer aficionados can use the storing and sorting facilities of this device, but unless the task is immense or a system already has been developed, the labour is not worthwhile for individual work.

Submitting letter

For a written communication there remains the matter of the submitting or supporting letter. There is a persisting and naïve belief that editors can be positively influenced. This is not so. Therefore, the type of letter which follows is counter-productive because the built-in resistance to being influenced favourably does not exclude a tendency to be negatively affected.

"Dear ...,

I enclose for you a manuscript of a paper on ....
This study is the one I spoke to you about and represents an entirely prospective study of ... in contradistinction to .... I think it is quite an important piece of work and you will see our figures compare very favourably with those recently reported by .... The work was presented at ... in ... earlier this year and was well received and provided much stimulating discussion.

It would be nice to see it published in ... particularly as you have
accepted the earlier work on the same subject.
I look forward to hearing from you about this.
   Very best wishes and personal regards,
            Yours sincerely,"

Equally, if one supports a submission by someone else one is likely to
evoke the cold shoulder. All that is needed to accompany a paper is a formal
letter of submission in the following style.

"The Editor, The ...
   Dear (Dr) (Mr) as the case may be. (Sir is rather formal and often
   implies that you don't know who the editor is. It is a courtesy to find
   out.)
   I enclose a manuscript on ... which I would be grateful if you would
   consider for publication in your journal."

Some, if not all, journals require an assurance that the work reported has
received the approval of an ethical committee or, if animals are involved,
has been carried out within the framework of the appropriate code of
practice. Nothing is lost and something may be gained by putting in a note
(not necessarily in the body of the paper) to this effect.

Finally, some journals need the signatures of all authors so as to protect
themselves from the possible accusation that one or more are publishing
without the agreement of others. Time may be saved if all those who appear
on the title page also sign the submitting letter, but this may be unduly
cumbersome.

Referees

With few exceptions, editors send papers to referees for an opinion. Therefore,
at least one person who is familiar with the field that the work covers (or
who thinks he is, or whom the editor thinks is) will assess it critically. In all
probability he will be a regular reviewer who has acquired experience, skill
and most importantly, in-built attitudes. This is a good reason for the writer
to try and find someone in his department or circle of acquaintances with a
background of refereeing and to ask him to read a draft. It is further wise
to accept his amendments, provided they do not distort logic or sense.

The referee system is potentially a threatening one. The academic world
abounds with stories of how referees have arranged to delay a paper so as
to repeat and publish under their own name the work it contains, or have
rejected a manuscript because it conflicts with their personal biases. Jones[7]
has pointed out that the absence of such abuse depends on the high moral

fibre of the referee. That this is not always present is also clear from his description of an instance of frank plagiarism after a submission by a British author to an overseas journal. I am sure the same thing has also happened in reverse.

Not much can be done about this situation except to hope that the editor one selects is a good judge of the ethical behaviour of his referees. What is not sufficiently recognised by those starting out to write papers is that nearly all journals with a referee system are prepared to provide the referee's report on a rejected manuscript. Usually, and for obvious reasons, editorial staff are not in a position to undertake lengthy and disputative correspondence. Nor at the moment are editors disposed to reveal names, though a little research into literary style and the nature of the criticisms will often create a short list or even indicate an individual. It is a pity that what Jones has called the "inviolate sacred cow of anonymity" should represent such a fixed point in the system. Anonymity obviously provides freedom of comment, but this can easily be confounded with absence of responsibility. There may also be a seniority conflict in relation to either a referee or an author. But these difficulties have been allowed to take precedence over the central fact that informed criticism is part of the framewor of all stages of the scientific process. Referees faced with the need to disclose themselves may adopt a more helpful attitude; some useful education could take place on both sides and might be continued if either as a rule or at the direction of an editor the author and referee were placed in contact. In that I am writing a simple primer on how communication in science can presently be done rather than a charter for the future, I shall not pursue the subject further here.

A final point that the author of any manuscript must try to remember is that although he has invested time, energy and emotion in it so that its importance to him is real, referees and editors have not the same personal commitment. A crisp rejection or even one studded with helpful criticism and advice may arouse in the writer pain out of proportion to the real injury administered. A cooling-off period is always desirable before the matter is hotly disputed; usually when the inflammation has settled the wound is found to have healed.

Returned and rejected manuscripts

When a manuscript is returned for revision and resubmission it is essential to deal with *all* the referee's (and *all* the editor's) comments in detail. Nothing is more likely to upset a critic than to find that something *he* regards as important has been ignored in a subsequent draft. To consider all the suggestions is not necessarily to accept them. Critics can be wrong. If, after due reflection, the author believes that he is right, he should feel at liberty to counter with his reasons when he resubmits. Obviously, the case made should be a good one, not "I don't agree" or "I cannot accept."

Rejection is more complex; Jones cites a rule that the likelihood of its occurrence may be in direct relationship to the ultimate importance of the work. Though I would not, on the basis of my own experience of both being rejected and rejecting others, go quite as far as this, I suggest that the author looks closely at the reason for rejection. If it is because the work is not regarded as either sound or important then personal re-examination or the opinion of a colleague should be sought. A fresh look of this kind may expose the unpleasant fact that there is less here than was thought. Then the hardest thing to do, but in my view the right one, is to refrain from trying elsewhere (and probably in what is, according to the professional jargon, a "less reputable" journal). The correct course of action when the cold light of reason suggests incompleteness, unimportance, irrelevance or negativity is to consign the manuscript to a bottom drawer and the idea to a mental pigeon hole. From these sites both may emerge in six months or so accompanied by fresh thoughts.

If rejection is because of unsuitability for the journal concerned, then a reassessment of the possible appropriate place is clearly in order (see Choice of Journal, p.6) - an unpalatable task, because differences in house styles will usually necessitate a complete new manuscript. There is an understandable tendency either to cut corners in its preparation or, even worse, to use a scissors to patch up a new version which is then photocopied, warts and all. Success with the next choice is then reduced.

Finally, if rejection is on account of length, prolixity, poor style or bad grammar - and, because many editors are prepared to scorn delights and live laborious days in correction if they consider the matter of a paper worthwhile, these are rare as single or combined causes - then this is the time to swallow pride, to recognise that skill in presentation is not given to all and to return to the literary workbench. What will follow later in this series are some of the tools that may help, including the advice not to write sentences as long and as full of conditionals as the preceding one.

Correspondence

If a paper is published in a journal with a correspondence column it may provoke letters. I am not at all certain what correspondence columns are for: we all enjoy reading them; they presumably act as a forum for debate, but unfortunately the exchanges can become acrimonious or merely posturing. Perhaps we have only ourselves to blame for not making the standard high enough and especially for our failure to examine why we write and how, as authors, we reply. James Learmonth made it easy for himself and his staff by insisting that the only two reasons for writing to a journal were to promote a charity or to defend a colleague; but this was literary Calvinism. Some slightly less confining rules can be made.

1. As a commentator on others. The situation is very similar to making a contribution to the discussion at a meeting (see p.64). Insufficient reasons under this heading are: paranoia at the failure of the author to mention the important contribution made; a desire to be seen in print; a need to inform others that this area has been better exploited by the letter-writer. Sufficient reasons are: correction of error if it is of clear importance to the underlying theme or if its perpetration may lead others into the same mistake (especially where patients are concerned); extension of the speculation which has been made by the original writer; refutation by additional information or reasoning.

2. As a reply to critics. Because criticism made on points elaborated rarely have weight or substance, it is seldom desirable to keep a correspondence going. However, from time to time some sort of reply is needed. Obviously it must be short, impeccable in its logic and, above all, courteous – the last difficult to achieve if the preceding letter has been unkind. A relaxed style and a readiness to admit error are most likely to bring favour both from the editor and professional colleagues. Soft answers do indeed turn away wrath, but this is not an argument for allowing a self-styled expert to ride rough-shod over good work with which he just happens not to agree.

REFERENCES

1. Calne, R.Y., and 11 others (1970) Hepatic allografts and xenografts in primates. *Lancet,* **1**, 103-106.
2. Nature (1970) Miscellaneous intelligence. **225**, 784.
3. Ahrens, E.H. (1974) Homozygous hypercholesterolaemia and the portocaval shunt. *Lancet,* **2**, 449-451.
4. Gawler, J., Bull, J.W.D., Du Bulay, G.H. and Marshall, J. (1974) Computer assisted tomography (EMI scanner). *Lancet,* **2**, 419-423.
5. Chagall, R.G. (1974) Building the tower of babbie. *Nature,* **248**, 776.
6. Watson, W.C. (1970) On psychological factors and disease states. *Gastroenterology,* **59**, 646-647.
7. Jones, R. (1974) Rights and wrongs of referees. *New Scientist,* **61**, 758-759.

4. English, particularly but not exclusively for written communication

Satisfactory communication by language embraces grammar, logic and style. Much is known about all three, but nevertheless it is difficult to extract from this mosaic of knowledge a simple pattern that will help the intending communicator. How also can we account for the fact that scientific communication is frequently regarded by a majority of critics as bad; or for the conflicting views held by many on what is the canon of "good English"; or for the very large number of books and articles (of which this is one) that seek to tell others how it should be done?

Nevertheless, we must face the problem. I shall divide what I have to say into the three areas already mentioned, recognising that there is a lot of overlap between them and that for scientific writing the first two may make the greater part of the third.

Grammar

We all have some of this embedded in our minds mainly as a series of conventions. Recently there has been a great deal of debate on whether or not there are rules of a general kind built into the basic connections of the brain and if so what these rules might be. The matter is a fascinating one but as yet seems to have little practical application for the man who wants to make himself clear. He either has, or acquires fairly early in life, intuitive understanding at a simple level of the ungrammatical nature of certain structures. This acquisition is reinforced during his school years by the imposition of certain conventions, some quasi-logical and some arbitrary. The first relate back to the clear meaning of sentences - such matters as the ordering and combining of subject, object, predicate, complement, and the association of particular sentences with particular classes of verbs. These rules are often language-bound - that is, what is acceptable in one language may be inadmissible in another. In that I am dealing only with English, this need not concern us. Further, it is usually possible as a consequence of long practice to write satisfactory and grammatical sentences without more than a subconscious appreciation of the general rules of syntax. A conventional ordering of words then results; easy communication follows because reader or listener finds the sentence structure familiar.

Trouble arises as expression passes imperceptibly from conventions which indubitably are part of the formal structure of English to the second class - those which have acquired an arbitrary status. This applies also to the meaning of individual words. The very fluidity of the language, which is part of its power, is such that purists who want to preserve certain patterns of words and certain specific meanings and intents of phrase are forever in retreat from the encroaching tide of common and convenient usage. As often as not a rearguard action to protect a particular situation proves to have been misguided in that the construct or word had only been temporarily frozen and circumstances are now transpiring to change it back to a former or more general application. All this - which can be found debated at length in any popular introduction to English usage - makes discussion about the *proper* use and combination of words sterile. Proper is always relative; what is more important is *consistency*. In any one piece of prose, words and phrases must always mean the same thing. If what they are supposed to mean is not what common use would expect or if the reader may fairly be assumed to be ignorant, then in addition they must be defined either as they are introduced or in a glossary. Should the latter method be used, the first time the word appears it should be signalled by a reference (footnote or bracket) to the glossary.

Care should be used in constructing a glossary so that it does not confound confusion. For example, an entry which refers to somewhere else, particularly if the reference is difficult to trace, would be less than helpful in that it sends the reader off on what could prove to be a wild-goose chase.

To seek consistency and precise definitions is a way of mental clarification. A check reading of a rough manuscript should endeavour to scan the text for ambiguities of this kind. The final draft should be written with a defined glossary at hand.

What is true of words is also true for arbitrary rules about word clusters such as split infinitives. Many of us were brought up to regard certain things as bad form and to condemn for example "different to" as an alternative to "different from". One can leaf through Fowler's *Modern English Usage*[1] without finding any defence for such purism, and though transgressions of this kind may jar they are relatively unimportant. Often they tell us more about the particular environment in which the writer was educated or the group with which he is trying to communicate. Nevertheless, there is something to be said in favour of holding to some conventions, however arbitrary they may be. If one aims, as one should, not to put barriers between writer and reader then the less there is which appears strange the better. Scientific argument is often complex and difficult. If the listener moves within a familiar verbal framework he will not waste time struggling with strange words or constructs and so expend his patience and intellectual energy. Moreover, though an editor is supposed to be above such things, a paper presented to him in a manner with which he is familiar may have a marginally greater chance of success. This applies also to the use of the given house style when writing with a particular journal in mind. I do not wish to suggest that a paper will be thrown out if its references are given in the wrong form, only that care in such matters advertises the writer as a careful and considerate person. It would be odd if this did not add a fraction of a grain to the weight of his acceptability.

Logic within and between sentences

The micrologic of sentences

Unfortunately it is not true that a rough intuitive understanding of the agreed rules of syntax always means logical clarity. Within a sentence there may be ambiguities and mis-assignments. Between sentences there may be a failure of the argument to flow in an orderly manner. To avoid these sins there are some general rules, though these are almost certainly incomplete in that the formal logic of English is still under study. The only way of reducing logical uncertainty is to submit all you write or say to a searching self-examination, taking it apart for both the meaning of sentences and the continuity of the thread of the argument. Aids to this process are to be found in the detailed analysis of Graves and Hodges, summarised in Table II, and taken from what, in my view, is by far the best available critique of effective logical writing.[2] My general advice attempts to illustrate some common faults, to enlarge upon their analysis and to apply it to scientific communication.

Macrologic

The foregoing paragraphs have dealt mainly, if not exclusively, with the structure of individual sentences, or at most the relationship between individual sentences - what I have called the micrologic of what is written. The composition of intrinsically clear and uncluttered sentences does not guarantee that the overall logic of the text will be satisfactory; particularly in the discussion section of many medical papers there is a tendency for the author to wander about, for example, from morbidity to mortality or from the inferences to be made from one set of data to those from another and back again. However, most writing is a method of displaying the processes of deductive inference and consequently should follow a logical pattern. In the composition of a paper it is well to pay attention to the

Table II Rules of style modified from Graves and Hodges

Every word and phrase should be appropriate to its context

Definition	Logic, Grammar
1. Every unfamiliar subject or concept should be clearly defined.	1. Ideas should not contradict one another or otherwise violate logic.
2. No word or phrase should permit of ambiguous interpretation.	2. No unintentional contrast between two ideas should be allowed to suggest itself.
3. In each list of people or things, all the words used should belong to the same categories of ideas.	3. All antitheses should be true ones. Overemphasis of the illogical sort tolerated in conversation should be avoided in prose.
4. No important detail should be omitted from any phrase, sentence or paragraph, and by contrast no unnecessary idea, phrase or word should be included.	4. No statement should be self-evident.
	5. The order of ideas in a sentence or paragraph should be such that the reader need not rearrange them in his mind.
5. It should always be made clear who is addressing whom on the subject of what.	6. Sentences should be linked together logically and intelligently.
6. There should never by any doubt as to *where,* or *when* something happened or is expected to happen or as to *how much,* or *how long* or *how many.*	7. Punctuation should be consistent and reinforce the orderly and logical presentation of ideas.
	8. No theme should be suddenly abandoned.
7. It should always be made clear which of two or more things already mentioned is being discussed.	9. No phrase should be allowed to raise expectations that are not fulfilled.
	10. Unless for rhetorical emphasis or necessary recapitulation no idea should be presented more than once in the same passage.
	11. The writer should not, without clear warning, change his standpoint in the course of a paragraph.

macrologic by blocking out headings for the discussion section or sometimes for the whole paper. Failure to do this produces a rambling, inferential style which makes the discussion too long and therefore invites rejection. Headings inside a discussion are permitted by some journals, but even if they are not they can implicitly be there provided the opening sentence of a paragraph is constructed as a lead-in to its subject matter. Informative headings or good leading sentences are likely to catch the eye of the reader; once caught, he may continue to read.

One of the important features of macrologic is the conjunctional link between sentences. One can overdo an analysis of these words, which are analogous to the symbols that link the steps in a series of mathematical or logical statements, but their consistent and accurate use does contribute to easy understanding. Spoken English has blurred the meaning of many of them, but in formal exposition it is helpful to have some convention to avoid ambiguity and to convey consistency. Here is one.

Thus and *therefore* are non-suppositional or unemotive links between steps in an argument that is usually deductive. Often they follow a semicolon or colon because of a need to imply a relationship of steps closer than separate sentences.

Because is stronger than *thus* and *therefore* in that it carries a sense of conceptual association (or aetiological in medical work), but has otherwise the same general meaning.

In consequence is a rather precious alternative to *because, thus* or *therefore.*

However, formally means in whatever way or to whatever extent. However, it is often used (as in the sense four words ago) as synonymous with *nevertheless* when it means "by contrast" or "in addition to". This is acceptable, I think, but Fowler discusses however in some detail and gives a slightly different view from mine.

Since should not be used as a logical conjunction because it is almost impossible to avoid its other meaning - "after in time" - at some point in one's writing. *Thus, therefore* and *because* are better.

As should not be used as a substitute for *because* for the same reason.

Following strictly means coming after in space (comparable to *since* in time), although this is pedantic and night does commonly follow day.

Albeit and *nonetheless* are elegant if slightly archaic variants for *however* and *nevertheless.*

Although and *though* imply contrast of two statements.

For is a weak alternative to *because.* It is acceptable in a series of mathematical statements, but less easy to use unambiguously in text.

Due to and *owing to* are other substitutes for *because* and are used as if they had the same meaning. Strictly, I believe that both mean repayment rather than a causative association and I avoid them. This may not be logical.

I have preached a rigorous approach to the use of logical conjunctions in scientific writing. We should not forget that less powerful links are sometimes appropriate to the spoken word or informal writing. "For", "yet", "of course" (the last to be used sparingly and to taste) obviously are good examples and can, before a critic points it out, be found occasionally in this text.

It is finally a pedant's pose to insist that these logical conjunctions, because they are analogous to links in a mathematical proof, should begin the sentence or clause which they govern. Overused, this convention makes for a wooden style. Nevertheless, if unchecked, a "however" or similar word may gradually drift down a sentence until it becomes wholly postfixed and consequently lose all its logical value. This is one of the few criticisms I have of the house style of *Scientific American* in which "howevers" frequently appear like apologetic logical white rabbits at the end of a sentence.

Classifications, both as a basis for headings or independently, can have value in formulating macrologic and guiding the reader or listener. However, here are a number of traps.

First, classifications must be consistent both in the categorisation of ideas (Table II) and in the indicators used. The following is a flagrant example of what can go wrong in relation to both.

"We conclude (1) Tachycardia may occur in the absence of fever and hypermetabolism; (b) hypermetabolism at normothermia occurs in the more severely ill (these patients are likely to benefit less from antipyretics); (3) during fever, these groups cannot be separated (d) separate factors need to be identified to account for the hypermetabolism and tachycardia of these conditions."[3]

It is mildly surprising that this passed authors, editors and proof readers, but having done so it provides an appropriate caricature. Fever, present or absent, is the predominant concept in the passage and the classification would have been more happily and more comprehensibly based on that. The confusion of the labels does not need further emphasis.

Second, classification of great depth may be a monument to industry rather than clarity. A taxonomic or hierarchical subdivision through more than three levels may be necessary for the archives but cannot be carried in the heads of an audience or reader and therefore defeats its own purpose for communication. Better in such circumstances to make a broad single or two level classification and within each segment to treat of the details in a different manner.

Third, labels can often add to the burden of comprehension. If the patients who receive antibiotics are called Group 1 and those who do not Group 2 and if thereafter these descriptions are used alone, the reader or listener must carry the definition which expresses the classification in his head. To do so is particularly difficult during a spoken paper or lecture. When a classification is necessary then some hint should be given in the label what this means (e.g. Ant. and No Ant. are preferable if less elegant than 1 and 2 in a paper where the use of antibiotics is being discussed).

Fourth, all classification contains an element of arbitrariness and at least implicit realisation of this should illumine the logic of what is written or said. Classification is a tool, not a be-all and end-all of understanding. The subject is a complex one and I shall not pursue it in any depth here. Suffice to say that classification must be based on a meaningful and a useful dichotomy.

Common faults

I shall now consider some particular faults often found in scientific writing and speaking. Examples have been chosen from the former because it is hard to get down in a reproducible way material from spoken communication. Nevertheless, I am sure that the same principles apply. A critical reader will say that the items of advice offered in relation to the examples have been frequently negated throughout the text. He would be right; I do not intend them to be more than advisory but they are applicable to the logical structure of scientific writing. They may be a little less relevant to a more relaxed expository style (see Brevity, below).

1 Complexity

Sentences become complex if they contain more than one factual clause and one generalisation. The reason is probably linear storage in the short-term memory of the brain. Unless there is some way of amalgamating clauses of fact, the sentence should be broken up in either of two ways: by a semicolon-colon arrangement, particularly if there is a need to group facts rather than to amalgamate them; or by simply having separate sentences for facts and generalisations. In the example below both things have been done. The result is longer than the original, but the gain in clarity justifies this.

> *Example:* However, the more discerning workers in tumour propagation did recognise tumour graft regression after a transient period of growth, and the host's subsequent refractoriness to rechallenge with the original tumour resulted from some kind of active repudiation of the grafted cells.

> *Alternative:* However, the more discerning workers in tumour propogation did recognise two things: the regression of tumour grafts after a transient period of growth; and the subsequent refractory state of the host when challenged again with the original tumour. Both these events could be explained by some kind of active repudiation of the grafted cells.

Another way in which sentences can come to contain too many clauses is by the successive, as distinct from the single, use of "and". The example taxes the short-term comprehension of the reader. This is particularly so if the clauses are disparate, as they are here.

> *Example:* The terms used in tissue transplantation are often ill-defined and have a multiplicity of origins and reviewing the literature it would appear that there are at least 20 words in current use to describe nine different types of graft.

> *Alternative:* Such is the lack of definition of terms used in tissue transplantation, and so multiple their origin, that a review of the literature suggests at least 20 words in current use to describe nine different kinds of graft.

The rewritten form has also taken advantage of the opportunity to put more emphasis on the whole message by adopting a semioratorical style. *Such* and *only* are the important words here because they provide emphasis. This technique can of course be overdone and therefore should be used only occasionally.

2 Related and particularly successive clauses or sentences should have the same structure

> *Example:* The sites of these excessive fibrin or fibrinogen breakdowns is usually not localised to one particular part of the body if it is intravascular in origin; on the other hand extravascular breakdown can also result in the production of large amounts of FDP even though the source of such fibrin is limited to one organ such as the kidney.

> *Alternative:* Intravascular breakdown of fibrin or fibrinogen is not usually localised to one particular part of the body; extravascular breakdown even at one site such as the kidney may produce large amounts of circulating FDP.

The writer was trapped by the "on the one hand ... on the other hand" construction. Such phrases are fillers which key the reader to expect statements of the same form. Here he has used half of such a statement only to confuse by a failure to make the matter after the semicolon conform in descriptive style to that which has gone before.

3 Undigested facts packed into sentences are difficult to process.

It is useful to aim at some synthesis of facts into concepts that can contain or extend the same message, or alternatively to separate facts from the concept by at least a semicolon. The first half of the following passage has been extended; the second reduced by doing some of the work of collation that otherwise falls to the reader.

Example: An enhancing effect on skin grafts has not frequently been obtained by pretreatment. Billingham *et al.* (1956b) used lyophilised tissue to prolong skin graft survival in mice which they considered to be enhancement without investigating antibody production: other possible mechanisms were also considered. Nelson (1961,1962) was able to transfer serum from guinea pigs pretreated with intraperitoneal spleen cells and obtain significant prolongation of skin graft survival in the second host; grafts on the pretreated animals also survived longer than the controls. Linder (1961,1962 a and b) was able to prolong skin graft survival by prior ovarian transplantation, the two tissues sharing the same antigenic determinants; the unresponsive state was preceded by a heightened resistance of grafting and was considered to be compatible with the findings of others regarding enhancement (Kaliss and Day, 1954). However, it was not possible to transfer the effect with serum nor demonstrate the presence of haemagglutinating or cytotoxic antibodies. Marshall (1962) using splenic cell suspension given intravenously obtained survival times of skin grafts in dogs twice that of controls, and preposed "desensitisation" as the mechanism, but did not detect any antibodies. Meyer and others (1965) used disrupted thoracic duct lymphocytes in rats given in total doses equivalent to 5×10^6 to 500×10^6 cells, orally, intraportally or intravenously in two separate injections followed by challenge with a skin graft one week later. The oral and intravenous route gave no significant prolongation or survival (range 9.3 - 11 days) but the intraportal route did give a significant effect. Depending on the dosage the mean survival times were 19.3 days following 50 - 70 x 10^6 cells and 31 days after 500 x 10^6 cells. The survival was also related to the cytotoxic antibody levels, the greater the number of dead cells in any dilution, the greater the length of survival of the graft.

Alternative: Pretreatment has only infrequently been associated with prolongation of skin graft survival. One example was the use of lyophilised tissues to lengthen the survival of skin grafts in mice (Billingham *et al.,* 1956b); although antibody production was not investigated and other mechanisms were considered, this was considered to be enhancement. The survival of skin grafts has been extended by transfer of serum from guinea pigs pretreated with intraperitoneal spleen cells (Nelson 1961; 1962), by prior ovarian transplantation in the rat when the skin and ovary share antigenic determinants, but not in the latter instance by transferring serum. Nor could haemagglutinating or cytotoxic antibodies be demonstrated (Linder, 1961; Linder, 1962 a and b). Prolongation of survival time of skin grafts has been achieved in dogs by the intravenous injection of spleen cells (Marshall, 1962) and in rats by intraportal, but not by oral or intravenous, administration of disrupted lymphocytes (Meyer *et al.,* 1965). The last was both dose-dependent and a function of the presence of cytotoxic antibodies - the greater the number of dead cells in any dilution the greater the length of survival of the graft.

4 Repetitive structure

Notwithstanding (2a) above which emphasises consistency of phrasing from the logical angle, a repetitive structure in style is boring and may make comprehension difficult. The example contained instances of this which have been eliminated in the rewrite.

5 Conditional statements and dependent clauses

Inevitably, much scientific writing contains complex arguments which are hedged about with conditional or contextual statements. Three things may happen:
   (a) The conditional clause may find itself in the wrong place.

   Example: If this thesis should prove to be valid then the data indicating that the surface membranes of isolated platelets contain clotting factors, such as Factor 11, after washing may reflect a pathological ...

   The "after washing" is tacked on as an afterthought when in fact it is a precondition of the other statements.

   Alternative: If this thesis should prove to be valid then the data indicating that the surface membranes of isolated washed platelets contain clotting factors (e.g. Factor 11) may reflect ...

   (b) The main statement is forgotten until the end of the text and becomes pendant to a number of conditionals or, as in the next example, on remarks which establish context.

Example: Gampton and Edlind (1966) have found that bile obtained from rats intoxicated with alcohol produces a more severe haemorrhagic necrosis than ordinary bile when injected into the pancreas.

Alternative: Injection of bile into the pancreas of the rat produces haemorrhagic necrosis which is more severe if the bile is obtained from animals intoxicated with alcohol (Gampton and Edlind 1966).

(c) Conditionals begin to make taxing demands on the reader's ability to scan the sentence.

Example: Some years ago we demonstrated (Fisher *et al.,* 1967) that plasma assay for a specific thrombin clottable fibrinogen derivative (fibrinogen first derivative of 265 000 molecular weight the first derivative formed from fibrinogen of 333 000 m.w. by plasma action (Fletcher *et al.,* 1966)) ...

Double parentheses - above - are always difficult to take in unless you are a computer programmer or (the same thing) an expert in symbolic logic. The contextual statements about molecular weight are stated differently and the double use of "first derivative" clutters up the sentence.

Alternative: Some years ago we demonstrated that plasma assay for a specific thrombin clottable fibrinogen derivative (fibrinogen first derivative (Fletcher *et al.,* 1966): molecular weight 265 000; formed directly from fibrinogen molecular weight 333 000) ...

6 Necessary but intrusive statements

Perhaps one of the commonest defects in the logical arrangement within sentences is to feel it necessary to include a statement but to get it in the wrong place. Such statements are supplementary rather than conditional; the effect is the same.

Example: We have studied the occurrence of thrombosis by the use of 125I fibrinogen scanning and at the same time the fibrinolytic activity of the blood by euglobulin lysis time and fibrin plate lysis very closely in collaboration with Dr ... in a series of patients undergoing major operations.

Here the necessary inclusion of Dr ... is misplaced. Furthermore, it is implied that without him the study would have been less rigorous, which may indeed have been the case but did not express the author's intention. A fresh start by a footnote, a parenthesis or, if the name is important a separate sentence are ways out of this difficulty.

7 Uninterpreted and/or complex mathematical statements

As mentioned earlier, figures on the whole are better out of the text. If they are included it should be clear *why*.

> *Example:* Twenty-six per cent had levels ranging 76 to 100 per cent; 39 per cent had levels from 51 to 75 per cent; 30 per cent had levels from 26 to 50 per cent and 5 per cent had inhibitor levels between 12 and 25 per cent of normal.

> *Alternative:* Only 26 per cent of women had levels at or near the normal range (76 - 100%). The majority (69%) were between 26 and 75% (39%, 51 - 75; 30%, 26 - 50) and 5% were in the 12 - 25% range.

In the alternative version the words *only* and *majority* and the separation of the last statement transmit the extra message that the distribution of figures is of a particular kind.

8 Its and this

Both these words, particularly when used in successive phrases or sentences, can result in illogicalities which may cause confusion.

> *Example:* The theory of active immunity is supported by the fact that an animal receiving a second allotransplant from the same donor will reject it more rapidly than it rejected the first.

The meaning may be intuitively clear, but the sentence says that the second transplant rejected the first. The first half of the sentence becomes retrospectively illogical. The reader may mentally do a double-take. If the argument is more complex than in this example (and it often is), chaos may reign.

> *Alternative:* The theory of active immunity is supported by the more rapid rejection of a second identical allotransplant made to the same recipient.

> *Example:* In this syndrome the primary event is intravascular fibrin formation. This is often accompanied by varying degrees of secondary increase in plasma fibrinolytic activity.

There is not much wrong with the logic here, though at first the reader may not be quite sure if "this" refers to the syndrome or to intravascular fibrin formation. However, "this" is unnecessary and is only there because the writer was too lazy to organise the logical structure in his head before writing it down.

Alternative: In this syndrome the primary event is intravascular fibrin formation which is often accompanied by ...

Yet another variant of this problem is the use of an inappropriate operator verb for a noun.

Example: Nevertheless, the principles described ...

Principles do not describe: they exist.

Alternative: The principles were .....

9 Introductory participles

Participles at the beginning of a sentence are always tricky. They tend to be misrelated, unrelated or (as in the example which follows) confused with adjectives (see Adjectival Ambiguity, below).

Example: Challenging mice lacking this antigen with sarcoma grafts bearing this antigen excited the formation of "isoantibody" (alloantibody) directed against it.

Alternative: Mice which lacked this antigen formed an "isoantibody" (alloantibody) when challenged with a sarcoma containing it.

The original could be read that the mice were either aggressive or challenged the experiment even though they lacked the antigen. Similar problems occur when a sentence begins "Arising from". The rewrite has precision although it implies rather than states that the isoantibody was elaborated in response to the antigen. However, in context this is an obvious implication.

10 Misrelation of statements

Example: A review of the mortality of acute pancreatitis from published series throughout the world reveals a wide scatter of figures.

The writer was reviewing published series to find the mortality, not trying to ascertain the effect of published series on the mortality. This is a trivial error in that meaning is clear; confusion can result in a more complicated argument. The alternative is preferable.

Alternative: A review of published series throughout the world reveals a wide range of mortality.

11 Adjectival ambiguity and misrelation

> *Example:* The evidence for alcohol producing spasm of the sphincter ...

The sentence can literally be taken to mean that spasm of the sphincter produces alcohol.

> *Alternative:* That alcohol produces spasm of the sphincter ...

What this example illustrates is emphasis. To use adjectives before a noun such as alcohol modifies the nature of the noun. To use them afterwards modifies the circumstances around the use of the noun. Here is another.

> *Example:* High flow oxygen.

What was meant was the alternative.

> *Alternative:* Oxygen at high flow.

12 Imprecise use of words and phrases

(a) Double meaning of words. I have already mentioned that there is little value in acting the semantic Canute when the waves of change in meaning are forever lapping at the shore of established use. Languages grow and decay. However, in scientific communication not only is it important to define precisely what a word means in a particular passage, but also it is proper to avoid words which can have more than one meaning in either common or scientific use. Often in practice ambiguity of understanding may not be involved; but if science is the removal of confusion and the precise investigation and description of the world, then it is our duty to be as accurate as we can. Examples which occur frequently in day-to-day medical communication are: "dye" when "contrast medium" is meant. "Bacteriology", "pathology", "radiology", etc. when what is meant are the bacteriological, pathological, or radiological findings. For example, this sentence -

> "There is no evidence of any long term alteration to small intestinal bacteriology after vagotomy and drainage procedures" -

debases the word bacteriology from its definition as the science of micro-organisms. The use of any word that defines a science or an area of knowledge as a synonym for a state, as in the above passage, is imprecise. It appears to be particularly common in the neighbourhood of London where "Do you see any pathology?" is a slipshod substitute for "Do you see any disease, abnormality, lesion?" Here is an example from further north.

> "At present the most sensible advice to any surgeon in genuine doubt as to the diagnosis is to proceed with laparotomy. Some other remediable pathology may exist."[4]

Although the most frequent offences are committed against the sciences, nouns which describe concepts or states may suffer the same transformation. Thus "even the passage of a little diarrhoea"[5] is inappropriate. It is possible to have diarrhoea, but you cannot pass a concept *per anum* even if, as here, it is little.

> "Even the passage of a small amount of loose stool ..."

would have been a more exact substitute.

As a final instance of the desirability of precision, words which have come to have a particular technical meaning in statistics should not be used in their common way unless the context is unequivocal. "Correlate" and "significant" are two good examples. "Incidence" and "frequency" are others.

Case is another much abused word in medical writing, having become a catch-all for "patient", "instance", "event", the formal description of an illness (case report), and as part of an operator phrase of this type:

> "In any case we do not know that measurements made in the general circulation reflect what is happening at the site of thrombus formation."

The last is also ambiguous in that it could be read either as "In any one instance ..." or (colloquially) "Anyway ...".

Case used to describe a person is a dehumanising word we, in a supposedly humane profession, should avoid. It is allowable to talk about "Mrs Smith's case" (i.e. her illness), but less proper about "the case of Mrs Smith", particularly in her presence. A good rule is to change the word for something more precise at every opportunity.

(b) Negative noun phrases (some readers will dislike that phrase, but it does accurately express what I mean). To criticise statements such as " ... with no history of diabetes" is to stand on shaky ground. The symbolic logician often legitimately uses 'not' to indicate the state of absence of some entity. Thus if R is red, $\bar{R}$ is not red, and forms a convenient manipulative tool of logic. The electronic engineer does the same with 'not' and 'nor' in circuit design. Yet expressions which involve having something which is not there do jar against common sense and may lead to statements which border on the ridiculous:

"No valves were visible on X-ray studies"

is comparable to the man met on the stair who was not there.
    Usually it is fairly simple to avoid this construction in formal writing and create a negative verb at the cost of a word or two.

"Did not have a history of diabetes."
"Valves were not seen on X-ray studies."

(c) Failure precisely to express concentration or levels. In general conversation we speak loosely, if conveniently, of the "blood glucose", "the serum sodium", "the urinary potassium" when we mean "... concentration" in the first two instances and either this or " ... content" in the third. It is even more precise, if pedantic, to say "the concentration of glucose in the blood" so avoiding the adjectival use of a noun. Certainly, failure to insert the appropriate quantifier in formal speech or writing is thoroughly imprecise.

13 Brevity

"Everything that is necessary and nothing that is unnecessary" said Theodore Kocher about surgical technique, and within limits this is true of scientific speaking and writing. Though overcondensed delivery, particularly in the spoken word, can make life difficult for the auditor or reader, so also can prolixity. Some of the advice tendered in previous paragraphs should help the techniques of conciseness without confusion. Short sentences with satisfactory internal logic; a general logical framework and a strict avoidance of ambiguity are perhaps the most important.
    Case histories are often overinformative. Obviously, facts which would have an importance to future analysts of the work may have to be included, but a careful eye is needed to exclude irrelevancies that add bulk, but not substance, to a package for which minimal weight is desired.

Example: "A man of 56 years presented in the Casualty Department on the 9th June 1973 complaining of severe abdominal pain. This pain commenced in the right iliac fossa 48 hours prior to his admission and had been steadily increasing." (39 words)

There is nothing overtly objectionable here unless one dislikes (as I do) the stylistic use of "commenced ... prior to" as an unwieldy substitute for "began ... before". However, do we need the repetition at the beginning of the second sentence? Are the dates relevant? Does it matter to the argument if the man came to the Casualty Department? The answer in this instance was "no" to all three questions and the edited version was:

> *Alternative:* "A man of 56 presented with severe abdominal pain which 48 hours before had begun in the right iliac fossa and had steadily increased." (25 words)

A trivial example, perhaps, but steady savings of this kind can cut the bulk of a paper by at least a third without interfering with either content or clarity. Also, economy can yield space for an extension when, in Calkins' words, "continuity and cadence merit the inclusion of a few extra words beyond those needed for cold logic."[6] A difficult discipline to learn is that of not writing to the limit in first drafts.[7] Initial economy should be a challenge; it is easier to expand from terseness than to truncate verbosity.

The cultivation of brevity is a courtesy towards the reader or listener, but a note of warning must be sounded. In speech particularly, there is a "one pass" situation where clarity must be such that the receiver picks up the whole message at once. Brevity permits more to be said, thus taxing the ability to absorb. Occasionally, therefore, it is wise to test the material to ensure that terseness has not been carried to the point of incomprehensibility. Rehearsal helps and should it emerge that there is too much "guid gear in sma' bulk", as the Scots would put it, some of the content may need to be removed and the draft eased open either by the insertion of more explanation or by the use of a more relaxed style. For the latter an analogy, a colourful phrase, or, very occasionally, the humorous aside may be useful. More often it will be desirable, if the passage is difficult, to repeat it in slightly different words, in summary form, or as a set of deductions. Alternatively again, the whole section should be recast so that a general statement is made and thereafter its dissection into parts allows the statement to be echoed as each segment is considered.

> *Example:* Statistical analysis endeavours to partition variability or, in technical terms, variance, between real differences in subpopulations and the random source of variation that can occur in any situation when samples are taken.

This is not an opaque passage (or at least I hope it is not, because I wrote it). However, it is quite a lot to take in by ear and warrants a continuation on these lines.

The simplest form, and that with which I will mainly deal, is the division of variance into two parts only, which we meet in statistical tests such as that of the differences between two means.

The element of repetition may be a helpful reverberation. Note also that the second clause ("and that" etc.) alerts the listener to the fact that what is to come will contain the main matters the author thinks are important. This form of explanatory filler can be useful in easing up an otherwise too tight or formal piece of writing or, as in this instance, speaking.

Rhetorical questions - which can of course be overdone and lead to an unpleasantly hortatory style - may serve the same function. Consider this:

While I say that adrenalectomy was only "a little more effective" than hypo-physectomy, what do I mean?[8]

The speaker (this was spoken communication) then went on to explain, but he had reminded his listener what the matter was about and where attention should now be concentrated.

Style

If consistent, inoffensive, conventional grammar, sentence micrologic and overall macrologic are all achieved in scientific writing, then style is a bonus which is all to the good but not essential. It can be argued that the basic elements of style are a consequence of the foregoing. Nevertheless, it is still possible to write inelegantly. We all do it when rushed or sometimes when a passage is overworked in an attempt at refinement. Some common failings are as follows.

1 Dictating machine style

Example: In the circumstances that these preoperative findings were considered to represent previous thrombotic occlusions, in this sick patient group, many suffering from carcinoma of the lung, these patients (in whom such prior thrombosis could not be detected by the use of 125I labelled fibrinogen) failed to contribute to the data.

Great skill is required to use a dictating machine for logically clear and elegant prose expression. Few can do it. The result is usually a string of conditional clauses. The alternative removes these.

Alternative: We believe that the preoperative findings were the result of thromboses which occurred in a group of ill patients, many suffering from lung carcinoma. Because 125I was not administered until the operation, we cannot compare the preoperative abnormalities with a postoperative diagnosis made by radioisotope scan.

2 Using fancy words and falling flat on one's face

Example: From the serendipity of a chance clinical observation we were able ...

He didn't understand the word and so the statement was redundant.

Alternatives: A chance clinical observation enabled us ...

or

Serendipity enabled us.

3 Vogue words and their misuse

Example 1: Because of these difficulties with streptokinase instead of using fibrinolytic assay methods for the clinical monitoring of this form of treatment, the thrombin time is used since it measures the amount of FDP released from fibrinogenolysis. A thrombin time twice that of the control value is considered a satisfactory parameter of this treatment.

Monitor and parameter are great in the right place. Neither is necessary here and both are incorrect.

Alternative: Because of these difficulties fibrinolytic assay methods are unsuitable for the clinical control of treatment with streptokinase; the thrombin time is used in that it measures the amount of FDP released from fibrinogenolysis. A thrombin time twice that of the control value is an indication of satisfactory treatment.

Example 2: The paracoagulation methods such as the protamine sulphate or the ethanol gelation tests have also been used to differentiate the FDPs derived from these two respective syndromes.

This is slightly more subtle. Differentiate is the vogue word and at first seems acceptable. Careful reading shows that technically it is wrong.

Alternative: The paracoagulation methods such as the protamine sulphate and the ethanol gelation test have also been used to distinguish between the FDPs which occur in the two syndromes.

4 Overwriting

Example: Fibrin proteolysis products released into a plasma milieu complex with plasma fibrinogen and are then detected as complexes of higher molecular weight than fibrinogen itself.

The writer was anxious to use *milieu* and *complex.* By doing so he made his meaning obscure.

Alternative: The products of fibrin proteolysis are released into plasma and become attached there to fibrinogen. The complexes produced have a higher molecular weight than fibrinogen.

5 Wearisome repetition

Example: There has long been a need for the development of blood assay methods capable of detecting the presence of an *in vivo* thrombus. Such need has now been met by the development of a technique termed plasma fibrinogen chromatography for quantification.

He was not thinking or he would have seen there was no need to repeat *need* and *development,* nor to substitute *technique* for *method.* Poor revision of one's own writing often contributes to this turgid style.

Alternative: There has long been a need for the development of blood assay methods capable of detecting the presence of an *in vivo* thrombus. Plasma fibrinogen chromatography fulfils this by quantification.

6 Padding (syn., fluffy prose)

Example: Consequently, important technical changes in methodology were needed described briefly later prior to the introduction of these promising methods to the clinical sphere.

The adjectives are unnecessary: *important* - how can it be defined; *promising* - who would work on the problem in relation to clinical use unless the methods were promising? *Methodology* is a word which can now be found even in the *Times Literary Supplement,* but it means the study of methods and their science and should not be used as a synonym for methods alone. *Clinical sphere* - this is blah blah. *Prior to* is pompous.

Alternative: Consequently, changes in methods (described briefly later) were needed before the technique could be used clinically.

7 Condescension

"The reader will be familiar with the non-equilibrium thermodynamic equations"; "summary analysis showed"; "it is easy to see"; "previous published work by the author". These are all examples of misplaced superiority: they imply that the reader ought to know and if not that he is at fault. Counter-attack may follow in which an assessor looks for some flaw in method or analysis that he would otherwise have ignored.

The "look how clever I am" syndrome is a variant on condescension. For example, "A translation by the author from the original Greek", when it is scarcely relevant to the thesis and has obviously been put in to show that it has been done, falls into this class.

8 Impersonal prose

It is almost an article of scientific faith that the first person singular or plural should not sully the pages of scientific communication. Why so? Perhaps because of the public and therefore assumed impersonal nature of science. Certainly, there is something to be said for not allowing the creator to intrude between the message and the reader or listener. Yet to believe that by adopting a style which appears to suppress self this will necessarily be so is to encourage a further misbelief that it is possible to be completely detached. Every worker is involved with his work; it expresses *his* interaction with a natural or contrived situation. Inevitably he is expressing himself when he records for public consumption.

Because of this built-in involvement, it follows that contortions so as to suppress direct expression are often overdone. To write of oneself as "the author" is lugubrious. To record that "the senior author did such and such" is pompous. To turn "we report here" into "it is reported here" is often unnecessary. Nevertheless, the overuse of the first person can both intrude and convey an impression of self-importance. It should be limited to situations where opinions are to be conveyed which directly relate to the individual or individuals concerned.

A further solecism is the use of the royal "we" when one person only is speaking or writing. It was a habit of ex-President Nixon - need I say more? One exception may be used sparingly. If the argument calls for an active participation of the reader (for example, the steps of a deductive process), it is permissible to make him feel that this is so by addressing him as, in effect, "you and I", which means the use of "we", as for example:

... if K is greater than 2 we may wish to examine the difference between particular pairs of means.[9]

9 Simplicity

It is often said by those, like me, who presume to offer advice on expression
that simplicity is a cornerstone of good English. Many of the examples of
error or difficulty that I have given contain this principle in that the writer,
by choosing a convoluted method of saying something, has become trapped
in the toils thereof. Most of the time the simpler things are the better;
but it can add to the spice of what is said or written, occasionally, in
Kai Lung's words, to "wander unchecked in the garden of bright images".
Here is Professor John Cohen reviewing an American book on the psychology
of death:

> "For the contents of these sepulchral and lugubrious pages will
> satiate the most voracious and omnivorous thanatophile."

Not simple, but beautifully put. A writer with a taste for or skill in such
fields need not be denied an occasional indulgence.

10 Saying what one means and meaning what one says

It is often difficult to disentangle *oratio obliqua,* frank evasion, jocosity
and the erosion of meaning that goes with overuse. Every amateur critic
builds up a collection of phrases which jar for one of these reasons and
which, when encountered, should immediately send out warning signals.
I have gathered some of them together in Table III. Many of them have
also become incorporated consciously or unconsciously into our methods
of expression in order to plaster cracks in a logical facade or fill holes in
an untidy or inadequate experimental design. At the evident risk of being
accused of whimsy, cynicism, or a combination of both, I list those that I
have found most entertaining and significant.

I have given but a few examples of the many stylistic habits which can
opacify. All of them set up a ground glass screen between the reader and
the meaning the writer intended to convey. They have been the most common
occurrences in my last few years of editing for two journals, but doubtless
many more variations could be quarried from the rubble of scientific prose.

Table III Phrases which should undergo critical appraisal

There are numbers of phrases that forever appear both in written and spoken papers and reports and grant applications. Before using one, or a similar form of speech in your own discipline, reflect if you really mean what you say or if another meaning is implied or concealed. Listen for them in the speech or writing of others and make the same analysis.

"It has long been known that ..."	I haven't bothered to look up the original reference.
"While it has not been possible to provide definite answers to these questions ..."	The experiments didn't work out, but I figured I could at least get a publication out of it.
"High purity ..." "Very high purity ...' 'Extremely high purity ...' 'Super purity ...'	Composition unknown except for the exaggerated claim of the suppliers.
"... accidentally strained during mounting."	... dropped on the floor.
"... handled with extreme care throughout the experiments."	... not dropped on the floor.
"It is clear that much additional work will be required before a complete understanding ..."	I don't understand it.
"Unfortunately, a quantitative theory to account for these effects has not been formulated by us."	Neither has anybody else.
"It is hoped that this work will stimulate further work in the field."	This paper isn't very good, but neither is any of the others on this miserable subject.
"Well known."	(1) I happen to know it. (2) Well known to some of us.
"Obvious", "of course."	Only by the most remarkable stretch of the imagination.
"In our experience."	We disagree with everyone else.
"The agreement with the predicted curve is excellent." '... good.' '... satisfactory.' '... fair.'	Fair Poor Doubtful Imaginary
"As good as could be expected considering the approximations made in the analysis."	Non-existent.
"Of great theoretical and practical importance."	Interesting to me.
"Three of the samples were chosen for detailed study."	The results of the others didn't make sense and were ignored.
"These results will be reported at a later date."	I might possibly get round to this sometime.
"Typical results are shown."	The best results are shown.

(Continued)

Table III (continued)

"Although some detail has been lost in reproduction, it is clear from the original micrograph that ..."	It is impossible to tell from the micrograph.
"It is suggested ..."	
"It may be believed ..."	I think.
"It may be that ..."	
"The most reliable values are those of Jones."	He was a student of mine.
"It is generally believed that ..."	A couple of other guys think so too.
"It might be argued that ..."	I have such a good answer to this objection that I shall now raise it.
"Correct within an order of magnitude."	Wrong.

(The above are extracted from *The Scientist Speculates*)

"We're taking this problem back to the laboratory."	We've made a mess of it.
"Surely it is clear to everyone."	Hopelessly obscure.
"We are making a pilot study."	We're taking a leap in the dark.
"We are heavily committed in this sphere."	We are chasing our tails.
"Interdisciplinary research."	Confusion.
"Cross-fertilisation of ideas."	Sterility.
"Research work in our own laboratory."	Unsuccessful efforts to catch up.
"Success rates of up to 80% have been observed."	Usually we achieve 40%.
"It is of course impossible to translate results from animal to man."	I am now going to do so.
"I am opposed to the underlying philosophy behind this concept."	I don't understand.
"Brown is doing a fine job."	No visible results.
"With great respect."	Totally without respect.
"A worthwhile project."	I think it is useless but I am afraid to say so.
"Breakthrough."	Doubtful clarification.
"We believe this experience is worth recording."	Negative results.
"Have you consulted a statistician?"	Your results are unpalatable to me.
"Further work is being pursued in this field."	If we can get our feet off the bench we will repeat the same *ad nauseam*.
"Fascinating."	Work by a member of my own group.
"Of doubtful significance."	Work by someone else.
"Unselected mongrel dogs were used."	Any half dead-or-alive cur was used.

(Continued)

Table III (continued)

"The curve has been fitted by eye."	A statistical technique would have failed to produce a fit.
"This slide may appear a little complicated."	This slide is quite incomprehensible.
"Disregard lines 4 and 7 on this slide and concentrate on lines 2 and 9."	I prepared this table for publication and thought it would do for display to ignorami like you.
"The audience will of course be familiar with the work of Blodsky."	None of you have heard of it but I want you to be aware that I have.
"Results (methods) previously reported from our laboratory."	Didn't we work hard last year.
"During the course of another study it was recognised that ..."	We are busy people but we retain a sharp eye that instantly recognises significant collateral information.
"It can easily be shown that ..."	Some mathematician friend derived this relationship for me. I couldn't.

(The above by courtesy of the Editor of *World Medicine*)

Theses

Composition and presentation

A thesis for a higher degree, or its near cousin an essay for a prize, has, or should have, four characteristics.

First, it aims to be the apotheosis of the author's work and intellectual achievement and thus should strive for the highest possible standard on subject matter, intellectual content, logical presentation and reasoned speculation.

Second, it must be self-sufficient; unlike the paper or review which usually directs the reader to refer to other sources for corroboratory evidence or for the follow-up of ideas, a thesis should offer the reader a complete account of its subject matter and should be readable as a whole and in isolation. A necessary corollary is that the review of the work of others must be (but regrettably rarely is) critical rather than an account of who did what. A thesis is a personal document, and while prejudice is not, at least in theory, part of science, value judgements undoubtedly are.

Third, if a thesis is to be evaluated as a whole, then it must contain all the raw data. The examiner can then trace the path and test the veracity of the deductions.

Finally, it must never be forgotten that a thesis is a work of exposition, which, within reason, should be able to transfer ideas to any reader who, while scientifically literate, is not necessarily specifically informed on the subject matter. It may be added that in most assessment systems it has *got* to achieve this aim, because there will be at least one uninformed person involved in its examination. Therefore, even more than in a paper, the argument must start from precise and consistent definitions (spelt out for *this* thesis in a glossary), must proceed from simple to complex and must not contain logical junps. To achieve this can be a wearisome task if the matter is long and complex. Nevertheless, it is essential because here as nowhere else the reader is an adversary. Mistakes of fact or of logic and errors of definition, spelling and printing will be sought with zeal. For example, an examiner may select every fifth or tenth reference and have it checked for accuracy; he may pursue a method by having an expert on his staff dissect it. The scientific world protects its own self-imposed standards with even greater assiduity than do the editors of journals, and the care required from the writer is thus even greater than when submitting a paper.

How can this be achieved? As with a paper, the thesis should be blocked out as a series of statements on content which also includes the logical steps required. As each section is written, its consistency should be cross-checked with the rest. Repetition as well as inconsistency can thus be avoided; also the opportunity can be taken to insert liberal cross-references between sections. As the work progresses the writer must repeatedly ask himself:

What am I trying to say at this point?

How does it fit logically with previous and successive statements?

Where am I now in the sequence of my general argument (or have I forgotten)?

Am I citing facts from others or do I need to be critical?

Am I cross-referencing enough?

Am I being repetitive?

Is it worth saying?

When a thesis of any size is composed it is relatively easy to keep track of the flow of ideas within sections, less easy to do so for the work as a whole. There is a terrible temptation to start production of the final manuscript before completion of a draft which can be read as a whole both by the author and by some critical colleague. This temptation must be resisted; it is impossible to gain an overview of a work by reading it in bits. In addition, glaring inaccuracies or omissions often become apparent only after the work is seen as a whole. Better to lose a few days or weeks in producing a coherent, readable draft *in toto* than to make the belated discovery that major excisional or reparative surgery is needed on a supposedly fair manuscript.

The exact structure of a thesis is often laid down by the university to which it is to be submitted. The form may at first appear unduly constrictive. However, provided the above questions are kept in mind, this need not be so. Furthermore, mild departures from the framework that has been proposed will not usually bother examiners provided the work is otherwise well presented.

Two special devices may commend themselves to the thesis writer. First, in a long and difficult argument, methods of making necessary but parenthetical statements should always be considered. A footnote may be useful and at least in a thesis does not bring forth the ire of publishers and printers who hate them. Alternatively, a proof or detailed justification for a method or an argument can be placed in an appendix. Second, when there are many illustrations, graphs, tables or a wealth of raw data, there is much to be said for submission in two volumes. The one contains the thesis proper; the other all the rest. The reader can then sit down with both in front of him and is saved the difficulty of referring from the text to an illustration two pages away or to a block of data at the end of the volume.

REFERENCES

1. Fowler, H.W. (1965) *A Dictionary of Modern English Usage.* 2nd ed., revised by Gowers, E. Oxford, Oxford University Press.
2. Graves, R. and Hodges, A. (1943) *The Reader over Your Shoulder.* London, Cape.
3. Halmagyi, D.J.F., Broell, J., Gump, F.E. and Kinney, J.M. (1974) Hypermetabolism in sepsis; correlation with fever and tachycardia. *Surgery, 75, 763-770.*
4. Irvine, W. (1974) Observations on acute pancreatitis. *Brit. J. Surg., 61, 539-544.*
5. Sykes, P.R. and Schofield, P.F. (1974) Post-operative small bowel obstruction. *Brit. J. Surg., 61, 594-600.*
6. Calkins, F.C. (1954) Word saving good and bad. *Science, 120, 614.*
7. Nature (1974) Editorial. It's your journal. 252, 337.
8. Atkins, H. (1974) The treatment of breast cancer. *Proc. Roy. Soc. Med., 67, 277-286.*
9. Armitage, P. (1973) *Statistical Methods in Medical Research.* Oxford, Blackwell.

5. Aspects of speaking at scientific meetings

Here I shall limit myself to one occasion - the presentation of a short scientific "paper" - because this is the bread and butter of scientific communication. The broader art of delivering an address, a lecture, a review or a speech (supposing the last can ever be classed as the communication of science) has been well and lucidly dealt with by others.[1]

I have put the word "paper" in inverted commas because I believe it is an inappropriate term. As I shall show, spoken communication differs in detail from a written paper although its basic framework may be the same. I sometimes wish I could find and propagate some alternative, but we are so indoctrinated to the phrase "reading a paper at such and such a meeting" when we mean "saying our piece" that it would now be a waste of time to invent a new word.

The historical background of the short spoken "paper" joins the obscurity of the development of scientific communication as a whole. Presumably it is th the outcome on the one hand of classical scholarly disputation and on the other hand of the early gatherings of the Royal Society. Be this as it may, the scientific meeting now has a well-developed form of which the short paper is a part. I am not going to defend or criticise this form here, though much could be written from both sides. Medawar, in a paper which I have already quoted, has covered a great deal of the ground.[2] Some other writers have also touched on the matter from various viewpoints.[3-6]

I accept the axioms that (a) there is some good in a meeting composed of short scientific papers and that (b) we must recognise that the word "paper" for such meetings has a different connotation from the written form. The latter, as we have already seen, is an ordered, highly stylised and often selective account. Usually, it makes light of difficulties which have been encountered, as though to admit that the work went wrong from time to time is to imply incompetence or that to boast of problems overcome is hubris. The level of intelligibility is geared to repetitive reading, the close scrutiny of complex but all-inclusive tables and often, at least for me, a technique of reading that includes a wet towel around the head and a high level of blood caffeine. The adoption of this dense (in the sense of tightly packed) method of transfer of information has become necessary because of the so-called knowledge explosion (better termed, perhaps, factual explosion). Scientists struggle with three competing trends: plenty to report;

58

editors who are desperate to conserve space; and the need to present information in such a way that the chain of inferences can be analysed by others. These condition the structure of the written paper. The relevance of this to speaking at meetings is that a paper presented verbally will not remain intelligible if it contains as much as a written communication or is composed in the same style as that for publication.

Nevertheless, so that a familiar pattern is presented to the listener and a coherent message may be transmitted, it is desirable, though not essential, that the spoken paper does not depart too far from that of its written counterpart. The same sequence of hypothesis (introduction), test (methods and results), inferences, modifications to hypothesis and occasionally speculation is appropriate. Properly done, this will permit the audience to follow the logic and, if need be, to challenge it.

Most of us have had more practice, if not training, in writing than in formal speaking. The result is that either speech is organised poorly so that recognisable form is absent, or the written style is transposed to the verbal. Neither of these gives a good product. What is meat for a journal may be indigestible for a lecture theatre. Organisation (and by this I mean what goes in and, to a lesser extent, where), style, and the use of cues which attract, hold or reactivate the audience's attention are all different for the spoken paper.

With this background in mind it is now possible to look in more detail at some of the essentials for successful presentation in papers that run at the most for 10 or 12 minutes.

Abstracts

A local secretary of the Surgical Research Society of Australasia recently sent out a request for abstracts with the following instructions:
1. The devil is come down unto ye, having great wrath, because he knoweth that he hath but a short time (Revelations 12:12).
2. The background, protocol and results of the study should be set out so that it is comprehensible to a trained monkey or to someone not working in the same field. Vague statements such as "the results will be presented and discussed in the light of modern knowledge" are unacceptable in an abstract. Unsatisfactory abstracts will be returned.

Both of these may be a little stark. Nevertheless, it is true that abstracts are often written in haste (and occasionally repented at leisure as the time of a meeting approaches). It is also much more likely that a committee which is vetting abstracts will respond to something all the members can understand and which carries the clear message of what the paper will both say and conclude. If it is impossible to write a clear abstract it will be equally impossible to deliver a clear paper, though the reverse is not necessarily true.

Writing an abstract is composing the summary of a paper which is in turn the paper in miniature - not a very profound statement, perhaps, but one departed from at peril. Introduction, methods, results and conclusion is the standard formula which will only rarely be departed from. Occasionally, two sets of results may require to be separated by a linking sentence or so to sustain the continuity of an argument. All the advice given for papers about the avoidance of ambiguity, careful, precise choice of words and the use of short sentences applies with equal or greater force to the composition of an abstract. In addition, there is here a good case for being as conformist as possible to the general form needed for the meeting. Tired referees going over a large number of abstracts will find it easiest if the texts are uniform.

Apart from this, abstracts are merely an expression of editorial skill and experience in the writer and his colleagues. As a guide to societies and others I quote in Appendix 3 the detailed instructions for abstracts put out by the Surgical Research Society of the United Kingdom.

Composition of the paper

The description I have already given for the structure of scientific inference in relation to communication is as appropriate here as elsewhere. The main objective of the paper, its *minimal* content to achieve this and the degree of additional speculation appropriate must first be clearly defined. Thereafter, some further points more germane to spoken than written work require consideration.

1. How much, if any, history or chronology is needed? A gentle introduction can often be achieved by telling an audience something simple from the past that 90 per cent of them already know. The majority will then split into two groups: those who congratulate themselves on being as well informed as the speaker and those who will now have grounds for hope that they may follow what is to come. The minority will have been given a starting point. Occasionally, but not often, a brief chrono- logical account will help, particularly if it contains a surprising emphasis - such as the neglect of the obvious over many years. However, the description must synthesise; there is nothing more dreary than "In 1945 Smith did this and then in 1956 Brown did that ...". Also, when intro- ducing what the speaker, in his wisdom, believes to be common knowledge it is dangerous to say "It is well known that ..." when in fact it is not at all. The conclusion that many listeners will reach is that one- up-manship is intended.

2. Is there someone in the audience who is working in the same field or has

made a previous contribution to it? If so, and provided it does not require verbal or conceptual gymnastics, it is gracious to refer to him in the introduction. Omitted, he will spend the rest of the delivery brooding; his comments at the end may well be directed not at the content of the paper (he has not listened) but towards establishing his claim to recognition.

3. How long should an introduction be? From experience with a stop watch during more than usually intense periods of boredom at some meetings I have attended, an arbitrary limit should be two minutes for a ten-minute paper - between 200 and 250 words of reasonably paced speech. If it is impossible to say in that compass what the work is about, then the whole structure of the proposed paper should be re-examined with a view to greater simplification. A long introduction leaves less room for a considered approach to other segments of the paper and particularly for a measured but stimulating discussion and summing up.

4. How much of methods should be included? Very little is the best answer. No one can take in a detailed account of a complex chemical extraction or the step-by-step description of a statistical analysis at one pass. Time is not available to be discursive and in the context of the short paper only two things are important. First, justification of the method by a reference in the abstract to the source of the technique. If the latter is new, or even if it is old but unfamiliar to the audience, a demonstration of its validity on a sample (separate from the model studied) may be appropriate and is likely to include some statement on accuracy and reproducibility as well.

Second, a description of the principles involved, particularly if these are likely to be new. Three ways of making this intelligible to the audience are encountered.

(a) A colour slide showing the way the apparatus is deployed; often accompanied by the best looking technician; usually not helpful unless it is essential to show how a patient is attached to or incorporated into the system.

(b) A schematic either of the apparatus (Fig. 2) or of the flow of an analysis (Fig. 3). In either case it may be useful to build up the scheme as the verbal description advances. For example, in the diagram in Figure 2 it could be broken up into three segments on successive slides (Fig. 4a, b and c). It should not be forgotten that visual images which change usually stimulate rather than sedate and that this technique also makes it less necessary to use a pointer.

(c) Finally, as already stated, it is impolite as well as impolitic to refer to a method by a code name (usually an investigator's; occasionally an eponym or acronym) which is likely to be strange to listeners. One-up-manship will again be suspected.

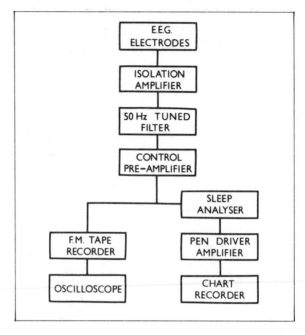

Fig. 2 Schematic of an apparatus.

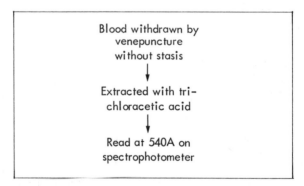

Fig. 3 Flow diagram of an analysis.

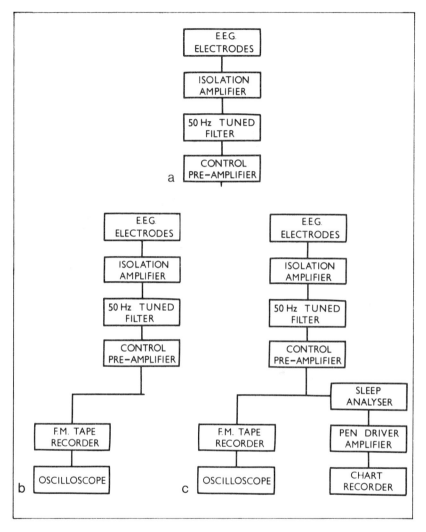

Fig. 4 a, b & c Method of breaking up an analysis so that it is sequentially
comprehensible.

5. How should results be presented? The most important error is a failure
to dwell long enough on each fact or groups of facts presented. Here
more than anywhere it is important that the slides are clear. There is
always the tendency to put too much into speech or visual aid and to
look for extra significance where it does not exist or, if it does exist,
where it is neither important nor relevant. Slides must be designed for
the specific purpose of transmitting or supplementing the verbal message.

Tables which may ultimately form part of a thesis or paper are nearly always quite unsuitable for visual presentation at relatively high speed. In the plan of what one is going to say, time must be allowed for the necessary reduction in speed of delivery that goes with the indication by pointer of key areas on a table or a graph.

The purpose of giving a paper is usually to point a conclusion derived from the data rather than to increase the audience's store of unanalysed or unsynthesised information. Therefore, the presentation of data should be specifically organised for this purpose: for example, a fitted regression line with confidence limits is more valuable than scattergram (provided always that the data in question stand up to such treatment and the statistical technique has not merely been used to idealise the situation). Not infrequently in non-quantitative work it is useful to show an example of what has happened (say, the production of an experimental ulcer and its prevention by a test agent) and, as the research work on which a paper is ultimately to be based progresses, it is also a good thing to bear in mind that at some subsequent date a photograph may be needed for this purpose. However, in selecting an illustrative example care must be taken to see that it is a typical one rather than the best.

6. What is the best form for discussions and conclusions? The general features of this section have already been considered. In presenting a paper verbally the principal need is to have the summing up precisely organised. Though slides which mimic what is being said should on the whole be avoided, there is a definite place here for a simple graphic or verbal display of the conclusions to give the audience a jumping off point for discussion. If suitably emphasised by a change in rate of delivery or tone of speech, a summary plus slide may even wake up those basically interested but *après déjeuner* dormant listeners in time to appear as if they had heard the whole paper.

The caution which leads to scientific orthodoxy at this stage has also been mentioned. In spoken communication discretion should surely be less than in the written word. However, the rule of once round the hypothesis-test circuit in a speculative way should rarely be modified, nor is there usually time for more.

Discussion and questions after a paper

These can be the best and most interesting part of the whole exercise. After all, one must assume that those men of the invisible College who eventually formed the Royal Society came together to get discussions going rather than to criticise the way one or other of them actually presented the material. Unfortunately, the concept that the *end* of a scientific paper is to provoke a discussion has tended to be forgotten in an excessive concentration on the *means,* which includes not only the methods used in obtaining the results but also the techniques of presentation. Such an attitude often leads on the one hand to a technical debate on methods which few can understand and on the other hand to silence because of the listeners' distaste for a bad presentation. If the presentation has been good then the second problem disappears; this in itself is an argument for a high standard of delivery. The first problem, that of questions directed towards method, can be avoided if the justification and limits of the techniques used have been clearly stated. Should such questions or debate crop up and if it is clear they are not likely to interest nine-tenths of the audience, it is reasonable to suggest subsequent discussion over a glass of beer or a cup of coffee. However, if the matter raised is of central importance to the argument then it should not be headed off. If justifications are not completely given in the body of a paper, perhaps for reasons of time, they should be available ready in mind or as a note so that a crisp riposte can be made.

What I have just written raises the whole question of anticipation. If some thought is given to what may be asked then semicoherent answers can be prepared; further, it is possible to have supplementary data in reserve. It is common practice to do this by having an extra slide. The technique is acceptable, provided there are not so many that the projectionist gets lost looking for the one wanted, or provided the speaker does not make it apparent from the fact that No.22 is asked for (after only six have been shown during the paper) that the situation is overcontrived to meet every eventuality. In order to save time and to ease the burden for the projectionist any numbering must be *very* clear.

The conscientious speaker at a meeting will have put a great deal of effort into his work and as far as possible attempted to eliminate errors, ambiguities and lack of logic. It may come as a shock to find that someone seeing it with a fresh eye may, like the referee of a written paper, deliver a devastating critical attack. The natural tendency is to adopt a defensive attitude or even to carry this a stage further into aggression. True, there are unworthy and thus irritating reasons which lead people to challenge in a discussion. Among them are: "I'm working in this area too and don't you forget it"; "I am going to reinterpret your work to show that my depth of understanding is greater than yours"; "You forgot me and my work in your introduction"; "I have an axe to grind"; "I want a grant in this field and there are grant-handlers in the audience"; and finally "I am an unpleasant, hypercritical, loud-mouthed individual who never misses an opportunity".

Nevertheless, all these having been accepted as manifestations of human frailty, there are still those who wish to discuss ideas because to do so is fun. Defensiveness or wrathful answers should not be for them. Rather it is desirable to be honest. If a point that has been overlooked, not read, or not properly analysed is brought up, the shortcoming is better admitted. The worst error is summarily to dismiss the comment as of no importance. If there is a good case for the criticism, a word of thanks to the individual who has provided it is a courtesy. However, empty catch-phrases such as "that's an interesting question" should be eschewed.

To be the subject of discussion and to be the discussant are but two faces of the same coin. I have already given some of the bad reasons for participation. Good or quasi-good ones are:

1. A desire to correct a mistake in fact or reasoning; but be sure the urge is linked to a reasoned counter-argument.
2. A feeling that there is a need to raise the general standard of what is going on; but be sure that you believe that this is possible.
3. A conviction that a worthwhile generalisation can be made; but do not have it emerge as a platitude.

Do not present a supplementary paper. Time for discussion is always limited. Some careful scrutiny of the abstract may help to focus down on a single point which is worth making. Further, no more than two separate questions should be posed. It is hard for both the audience and the speaker to retain each, particularly if the sentences used contain conditional statements

Coming to the rescue

A good communication, well presented, may fail to elicit any discussion. There are many reasons. First, the subject may have been well delineated and presented and no more needs to be said. Second, it may be too much outside the mainstream of the audience's understanding; not much can be done about that problem. However, there are situations when the discussion falters for want of an idea or comment which is relevant. Should a gap be anticipated a prediscussion can be held with a colleague working in the same field who understands what is being done and so can pick up a point by previous arrangement. Perhaps an artificial approach, but it is surprising how often with such a catalyst discussion which of itself is inhibited can be got going.

A careful reading of well-prepared abstracts can also help. As much information as possible should go into abstracts so that they can permit questions to be generated before the meeting. Perhaps this is coming dangerously close to precirculation of papers which, though common in the U.S., I personally deplore (see below). In the Departments of Surgery at

Monash and St. Mary's we have always gone over the abstracts and provisionally assigned to individuals those papers on which we feel able to comment. By this we endeavour to be assured that we will give of our best to the discussion. A military approach maybe, but one that is designed to make the most of a meeting.

At the time of the meeting it is occasionally a good thing if one person can act as a co-ordinator of a group's contribution to an individual discussion. A careful bit of eye-catching and a nod are usually enough to bring in the appropriate speaker so that only a few are then trying to attract the chairman's attention.

Precirculation of papers

As I have said, this is a common habit in the U.S. and less common in the U.K. The argument is that scientific colloquy can be better guaranteed if it is done. Judging by the discussions that follow, the only result is to give the commentator a chance to mobilise his experiential resources and convince others that in the particular field he also has something to communicate. At least in theory it should be possible for a well-constructed abstract to achieve the same, while keeping something up the sleeve which will preserve the spontaneity and thus possibly the creativity of the occasion.

Style

As with written work this word covers three areas - the appropriate choice and composition of words into intelligible sentences; the even and clear flow of logic; and the manner by which communication is achieved. The first two differ not a whit from written work and have already been dealt with. All that need be said is that even greater emphasis may be required in speech on the simplicity of sentences, the use of some repetition and the avoidance of purple prose.

The manner of delivery is a separate subject. Calnan and Barabas have dealt with it comprehensively, light-heartedly and well in relation to lecturing.[1] Their analysis includes clarity; appropriate rate and rhythm; the modulation of the voice; emphasis; and finally some indefinable factors which they call the delivery "style" and which I would associate also with the behaviour of the speaker. Table IV summarises their advice.

To it I add certain specific practical points as follows:

1. In most communication the focal points are three - audience, speaker and screen. By definition, and except in those awful meetings where people seem free to come and go as in Parliament or like travellers at a railway station, the audience is fixed - glued to the edge of their seats, it is hoped.

Table IV (modified from Calnan and Barabas[1])

Desideratum	Requirement
Clarity	Clear diction. Familiar and simple language. Simplicity.
Rate and rhythm	Less than 150 words a minute, preferably no more than 100-120. Slower for important passages. Variation for emphasis.
Modulation	Varied pitch, tone and volume for emphasis. Volume adapted to acoustics and amplification equipment (a precheck is vital).
Emphasis	Voice inflection for indicating important words or phrases. Restatement. Variety. Accompanying gestures.

For them, it is distracting to have moving targets; a jerky pointer and a jerky speaker are intolerable. Many of us have difficulty in standing still and also some (including myself) have trouble fixing any individual in the meeting with the eye so as to establish a point of communication and thus to avoid the wandering of our own eyes and feet. Point fixation is a valuable skill which should be acquired early.

2. The old theatrical rule of not turning the back to the audience should be studiously observed. To do so may require quick thinking when the venue turns out to have (as it should not) a point of delivery directly under the screen. When such an event takes place it becomes difficult to point at a slide without turning round or going into upright opisthotonos. A different form of words must be used as a substitute.

3. Although we go to a meeting to "read a paper" most societies specify that a paper should *not* be read. The reasons for this are obvious - reading encourages a head-down-dry-as-dust form of delivery and a prose style more suited to the written than the spoken word. It is a counsel of perfection that what is to be said should always be sufficiently organised and compressed so as to go on one or two cue cards, but to strive for this is worthwhile, even if some lack the skill to achieve it in the end. To use the manuscript as a cue is possible but hazardous, because it is likely that the place will be lost and a frantic turning over of pages may then ensue. As a way round the problem of getting as near word perfect as possible, some speakers use slides as cues so that what is about to be said appears almost identically on the screen. This is slipshod and an insult to the audience.

4. Humour has a small but important place in short scientific communications. A lot of scientific work has its funny side and a well-chosen remark can cheer up a dull meeting. I regard it as a mistake to capitalise on something that has happened earlier in the same meeting.

5. By contrast to this advice, if material has been presented in previous papers that has a bearing on the work to come, it is evidence only of a mental rigidity to avoid, as many do, any mention of that which has gone before. Urgent revision may be required to the structure of the paper so that gracious reference can be made or the evidence incorporated into the argument.

6. The sentences in a paper should not be strung together by a single repetitive phrase for summoning the next slide (the common one is "the next slide shows"). Elegant variations and phrases which also act as links in the argument, such as "turning now to", "if we consider", "combining, integrating, correlating the results leads to ...", are more relevant and stimulate jaded attention. Occasionally, when a projectionist is available, reversion to "may I have the next slide" ("can I see the next slide" is the in-phrase at the moment, but has an element of uncertainty about it) accompanied by a change of tone or emphasis will also bring back both audience and projectionist from the Land of Nod.

7. Projectionists, like pianists in the old time American West, are nearly always doing their best. They should not be asked to perform miracles of interpretation of requirements for an individual paper which to them is only one of 20 equally complicated or dull. A request to backtrack in a sequence should *never* be made; an extra slide should be used instead. If, in spite of these precautions, things go wrong, it is 10 to 1 against it being the projectionist's fault. Slide sequences should be foolproof and can often be shown not to have this quality by being projected by a departmental chairman at a rehearsal. In calling for a slide it is essential to alter the voice and rate of speaking. Few projectionists have sufficient technical knowledge to follow the argument and thus will lower their level of attention so that a buried "next slide" may pass unnoticed.

Rehearsals

Each intending speaker will have his own method of preliminary rehearsal: Demosthenes-like to a tape recorder rather than the waves; as Hamlet in an empty lecture theatre; or at the fireside to a patient wife. However, all roads should converge on a final series in the agora of the organisation to which the speaker belongs. If more than one paper is to be presented, rehearsal should include all involved so that a sense of egalitarianism and a corporate spirit are preserved. Everyone should be rehearsed, however junior or senior, and the rehearsal must take place in time to alter slides that are suddenly found wrong or inadequate. The use of an overhead projector to draft slides for a preliminary rehearsal will usually eliminate the need for this, but it is surprising how, every so often, a spelling error turns up that no one has seen. The objectives of group rehearsal are:

1. To get timing right (using a stop watch) under real conditions - that is, with slides and the steps necessary to explain them.
2. To check and amend the technique of delivery.
3. To work out what questions are likely to be asked in the discussion period.
4. To confirm or instil confidence.

The clash of ideas generated at rehearsals has been for me one of the most enjoyable features of presenting short scientific papers, although I do recall, in a less democratically organised department, being terrified by my Chief. This can hardly happen today.

Slides

Some years ago an excellent guide to the design of slides was published by the Department of Audio-Visual Aids at the University of Melbourne. Soon thereafter much of the information it contained was included in the instructions sent to speakers before a meeting of the Australasian College of Surgeons. Yet the first speaker, who was a member of a university department, projected a slide which broke nearly every rule that the guide had put forward. It can only be concluded that in this area there will always be those who are loath to conform, lazy, indifferent to the needs of others, or sometimes plain stupid. Since this incident many years have gone past, but recently I attended a meeting at which less than a third of the slides fulfilled the basic requirements of easy communication; one speaker uttered a memorable "You can't see this slide; I can and this helps me". However, for those who would decry my derogatory attitude in themselves, the following principles are offered. They lean heavily on the source already quoted.

Content

In general a table, diagram, or graph designed for publication and with an eye on the editor's desire for economy has too high a "packing density" for use in verbal communication. For this purpose a table should not be larger than two columns and two rows; and a graph should hardly ever include more than two variables. A diagram must be reduced to the absolute minimum required to illustrate those points to which reference is to be made by the speaker. If it seems that any of these rules is going to be broken then the text and draft slides should be reviewed to see if two or more of the latter are needed in place of one.

Graphs

Presentation of information by graphs is useful in that relationships and progressions (for example, with time or biological change) can be easily conveyed to the listener. A graph is indeed synthesised information and thus enables a concept to be communicated in an economical way. To achieve this in practice, certain conventions should be observed.

(a) The clarity of the message is more important than the absolute accuracy of the tracing. This is not a licence for fitting curves by eye or taking other liberties, but rather an encouragement to make the graph lines visually bold which can be achieved by having the lines thicker than the ordinates. A convenient rule is that the latter should be half the thickness of the descenders of the type face used, the graph lines being the same as the type.

A further point which, to my mind, increases clarity is to break lines at points where data are entered on co-ordinates. The principle is illustrated in Figure 5 and is particularly important on those rare occasions when two variables are plotted and require separate symbols. Lines that cross should also be broken (Fig. 6).

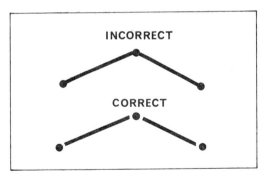

Fig. 5 Breaking a line at data entry points.

(b) The spacing of values on the X and Y axes should be as even as possible to produce a symmetrical graph on the slide. There is a danger in this - spreading the Y axis may exaggerate the visual impression given and thus overemphasise a relationship - but proper statistical analysis of the data should prevent this lapse.

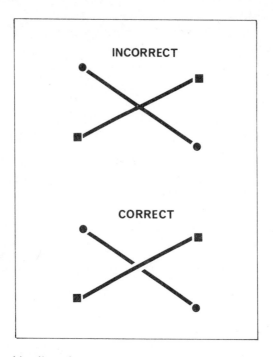

Fig. 6 Breaking lines that cross.

(c) Time scales are conventionally on the X axis.
(d) Values for the dimensions on the X and Y axes should be stated as clearly and as minimally as possible. Although dimensional analyses are useful and may be an integral part of a formal written presentation, they are less suitable for rapid visual assimilation and therefore should be avoided unless a special point is going to be made about them in the verbal matter. S.I. conventions in 1975 permit the use of a slash, which is probably more easily recognised. Do not succumb to the temptation to write vertically *up* the Y axis. This requires most people to lie on their sides.
(e) Extrapolation of graph lines beyond the limit of data available should be avoided because it is logically untenable except as a hypothesis - a matter particularly true of a regression analysis.
(f) Cumulative percentage technique (Cusums) are relatively unfamiliar to biologists except those who work in epidemiology. They often lead to squashing of data at a low frequency where a percentage change may be quite important but is obscured.

(g) Purists would maintain that a graph is a map of a relationship between variables which should have some algebraic expression independent of visual communication. Thus, a graphical representation for a biological variation such as that shown in Figure 7 for change in concentration against time may be inappropriate. Yet it is thoroughly convenient to do so and no harm results provided a formal mathematical relationship

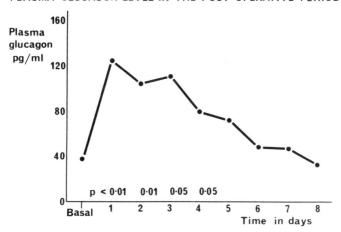

PLASMA GLUCAGON LEVEL IN THE POST-OPERATIVE PERIOD

Fig. 7 A change in concentration against time.
Strictly this is *not* a graph but the concept of biological variation is nevertheless well illustrated.

is not implied. Anyone who sets out to draw a graph of this kind must avoid the assumption that in joining two data points by a straight line or smooth curve - particularly at maxima or minima - he is necessarily making a valid intrapolation.

I am less happy about the use of graphical representation for plots of spatial distribution of data (Fig. 8). All the drawbacks listed above apply and there is even less logical reason than in a time relationship to assume that some form of continuity exists from one data point to another. It is better to plot the results as frequency distributions (Fig. 9).

(h) The indication of statistical limits on a graph may be essential to comprehension. When it is done care should be taken to ensure that the correct message is conveyed. For example, to indicate a standard deviation implies that a *range* is being displayed. A standard error is the degree of confidence that can be placed in a mean. Standard errors often look more attractive but are not necessarily appropriate to the task. Anyone using these methods of display should be conscious of the difference.

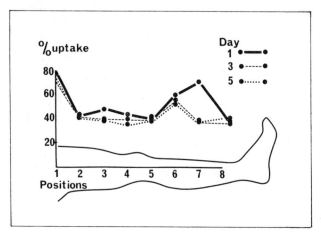

Fig. 8 The spatial distribution of increased counts has been plotted as a
graph. While legible, this is not correct. The distribution is wholly
discontinuous.

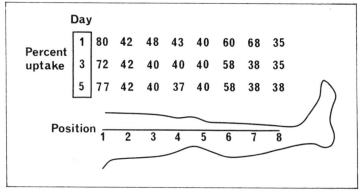

Fig. 9 More correct but admittedly at first sight less legible view of
Figure 8.

Histograms

These are very useful for expressing frequency relationships and can be
plotted vertically or horizontally, the latter (a bar diagram) being
particularly useful if writing is required (Fig. 10). I have difficulty in
taking in histograms at first sight or at high speed unless they are unequivoca
Two points follow (at least from my visual inefficiency):

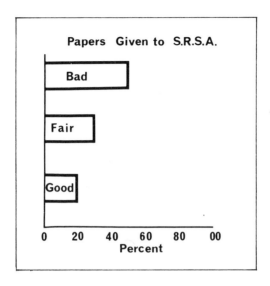

Fig. 10 A useful bar diagram.

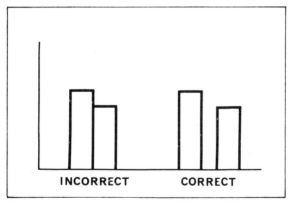

Fig. 11 Separation of the vertical bars of a histogram *may* make it more legible.

1. Black on white (or positive on negative) is the best convention; diazo (see later) is less suitable.
2. The vertical bars of comparative histograms are sometimes better separated (Fig. 11). This may require an extra slide or two.
   All the advice about charts and graphs given here can be found in extended form in a paper by Simmonds.[7]

Tables

The general rule about too much information has already been stated almost *ad nauseam.* It remains to detail their construction so as to achieve good communication.

(a) Blank space within a table makes it difficult to relate the figures to each other. The information should be centered round a clear margin. A table which will be presented as a slide need not be - the margin of the visual field creates the box. Indeed, the presence of a box (and often a tell-tale "Fig. so and so") may reveal that the speaker has lazily copied from a written communication without consideration of the needs of his audience. Finally, a box restricts the size of the material and thus also limits resolving power.

(b) Headings are not often needed. The heading should come across in what the speaker is saying. If it is felt vital to remind the audience of what has been introduced the heading should be single-spaced and separated by two lines from the content. The shorter a heading the better - less than a line if possible. If a second line proves necessary the typescript should be centered below the first.

(c) Figure 12 illustrates a badly constructed table by all counts and Figure 13 the correct version. Obviously the content of the slide should be centered on the heading. I have a penchant for centering successive

COUNCIL OF ROYAL COLLEGE OF SURGEONS
OF ENGLAND

	Mean Age	S.D.
Sitting Members	63.6	4.57
Candidates	56.1	4.33

n_1=25 n_2 = 24 F = 1.109 p <0.05

Fig. 12 A poorly constructed table - see text.

COUNCIL OF ROYAL COLLEGE OF SURGEONS OF ENGLAND

	Mean Age	S.D.
Sitting Members	63.6	4.57
Candidates	56.1	4.33

$n_1 = 25$ $n_2 = 24$ $F = 1.109$ $p < 0.05$

Fig. 13 Correct version of Figure 12.

MODEL

Small intestine of dog.

Each animal its own control.

One layer seromuscular anastomosis.

Fig. 14 Centered successive statements.

statements underneath each other (Fig. 14), but this is not strictly necessary and secretaries do not like it because it is tricky. Numbered statements obviously are best directly underneath each other (Fig. 15).

(d) The presentation of figures in tables should pay due attention to ease of interpretation and significant digits. For the first, when the number is less than 1, 0 should be inserted before the decimal point. Whole numbers should be followed by the appropriate zeros - e.g. 7.65, 8.00. For the second, the question should be asked "What are the significant figures?" There is all too often a tendency to overdo apparent significance. Bracketed figures which make conversions or additional statements (e.g. deaths or complications) should be avoided in a slide because they are hard to take in and force the watcher to do calculations which, if

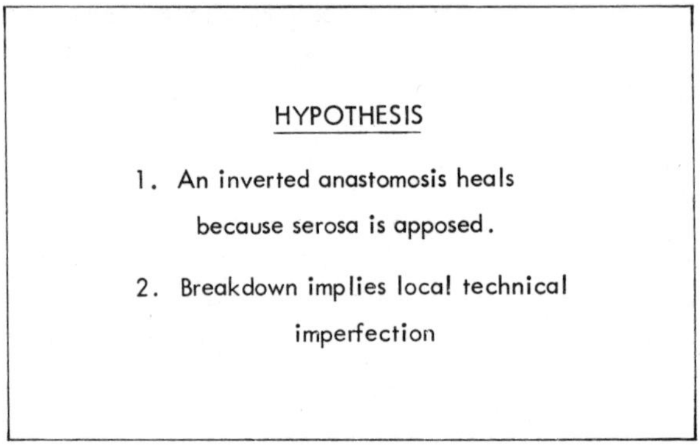

HYPOTHESIS

1. An inverted anastomosis heals
   because serosa is apposed.

2. Breakdown implies local technical
   imperfection

Fig. 15 Appropriately arranged *numbered* statements.

the point is to be made, are the responsibility of the speaker. Nil entries are best indicated by a dash, otherwise line scanning is difficult.

(e) A brief indication of a statistical test usually should be given at the foot of the table to save time being wasted in unnecessary discussion.

(f) A niggling point, but I think a valid one, is that care should be taken in rounding gross figures of less than 100 to percentages. Because one of two results turned out in a particular way it is not easy to say that 50 per cent will. If 35 of 70 did, it is much more likely that 50 out of 100 will. Common sense and context are the best guides.

Line drawings and photographs

In short scientific communications these are most commonly found in the section on methods. Their number should be held to a minimum and their contents simplified and formalised. This almost always eliminates a photograph, but if one is included attention to detail is vital: a dirty smudge on the bench, a bloody swab in the operating field, an uncut thumb nail all shout incompetence from the screen.

Line drawings are more easily interpreted if positive - that is, the background is less contrasty than the line. Obviously, this is how we see. Thus, white on black or diazo are inappropriate for line drawings and, in particular, the recent widespread use of diazo techniques for this purpose is a mistake.

Gimmick slides

There seems to be an increasing tendency to use joke slides in short papers.

There is nothing wrong with this provided the situation is not forced. Nevertheless, the practice can be overdone and one should always ask the question: Is it necessary? What am I achieving? Do not forget that the text of a cartoon will have to be read and this may distract the audience from what is being said.

Overdecorated slides

Tables against a background of the hills of your home country or symbolic representations of what you are talking about are quite unnecessary. Similarly, the use of many colours and types often only confuses. I have a feeling that here both audio-visual departments and research groups are in competition to outdo each other.

Ensuring visibility; resolving power and format

The characteristics of the human eye limit our ability to resolve images at a distance. Lecture theatres which have been designed with this in mind (and such is true in most newly built educational institutions, but less commonly the case in hotels or conference centres) aim to have the furthest spectator seated at a distance no more than six (optimum) or eight (maximum) times the longer dimension of the screen. For example, if the screen is 1.5 x 3.0m then the depth of the lecture theatre should not exceed 18 - 24m. This rule can be applied only if the material to be projected fills the slide and therefore the screen (a function of the hall and the projection equipment). Furthermore, these limits can only be approached under ideal conditions which are not often achieved because the image brightness of the screen, the efficiency of the projector and the level of ambient light do not usually favour the slide.

By contrast to these factors, the material chosen for a particular purpose, the format and the standards of drafting have a significant if largely subjective effect on visibility and thus on comprehension of a slide that is basically visible and is well projected in a suitable lecture theatre. These matters are under the control of the speaker and I shall now summarise them.

Types of slide

The three types of slide in common use are black on white (positive transparency), white on black (negative transparency), and diazo (negative transparency; background from a choice of colours the most common of which is blue).

Black on white is the oldest and still the most widely used. The image is usually frank and clear with enough background light independent of the subtleties of house illumination for notes to be taken. The chief disadvantage of a black-on-white transparency is that any imperfection in the background stands out very clearly and cannot be removed. The intensity of the background can be tiring and, if lines are thin, visibility may be reduced.

White on black is merely the negative produced by the first photographic process. Contrast is usually good provided type or lines are bold. Imperfections in the background can usually be "spotted" out. The auditorium will be too dark for note-taking unless special dimming arrangements are available. As already mentioned, the negative image is inappropriate for line drawings.

A refinement that can be introduced on a negative image is to colour the lines or type with a special crayon applied to the transparency. This technique is simple, but, because it is, it sometimes encourages a rainbow approach so that instead of being clearer a graph or table becomes more confused.

Diazo. The soft-toned background of diazo slides is very restful. There is a relatively high contrast without glare and the level of ambient illumination, as with black on white, is usually satisfactory for note-taking. The background is homogeneous and, provided the technique is good, free from flaws. Diazo slides are particularly useful where graphs and tables are to be mixed with colour transparencies because the intensity is roughly the same and jerky changes in the visual flow are avoided. Their drawbacks are: inapplicability to line drawings; lack of universal availability; some variation in background intensity from batch to batch; fading with age, though this can be corrected by reprocessing; increasing costs associated with the process. Diazo slides became something of a status symbol in the 1960s and early 1970s, but there are signs that their popularity is waning, perhaps chiefly on the grounds of cost.

The choice of slide is thus a matter of objectives and availability. One general rule is to avoid mixing slides; apart from making the audience adapt to different intensities and visual conventions (e.g. a switch from positive to negative) an impression - which may well be true - is conveyed that the slides used have been thrown together for the paper from a dusty heap on the desk, rather than that they have been prepared with the consideration of the audience in mind.

Format

The production of a department is best limited to as few standard formats as possible. Although in theory, and I suppose in practice in some well-funded places, departments of audio-visual communication undertake drafting, one usually finds that the task devolves on the speaker or the technician. To have a ready-made set of rough rules and the appropriate types is then convenient, particularly when a rehearsal reveals the need for an extra slide.

Two convenient formats are:

9 x 12.5cm with 12 point type (usual typewriter face), lines not less than 1mm. Work produced in this format should be easily visible to the naked eye at 1m.

21 x 29.5cm (standard A4 paper). Type not less than 24 and preferably 30 point. Lines not less than 1.5 and preferably 2-3.5mm. Should be easily legible at 3m.

If these formats cannot be used then the rules for non-standard ones are that all work should be in a height-width ratio of 2 to 3 and symbol size must be adjusted accordingly. Symbol height can be determined from formulae and most audio-visual departments have their own. I have not found these very useful for the amateur and prefer the practical test of viewing the work from a distance of six times the longer dimension. A final check on legibility is that the slide itself can be read with the naked eye,[1] but this is *post hoc* and therefore not very practical.

Drafting techniques

In spite of improvements in the electric typewriter and its wide use, the typescript is still inferior to pressure adhesive lettering in the quality of work it produces for tables and the letters on graphs. However, the labour involved is much less. Also, a secretary can rapidly correct or clean up blemishes on the draft for a typewritten slide by using the new whiteners. Thus, a good electric machine will be used for most tables or statements. Graphs come out much better if pressure lettering is used because the script then matches the lines according to the rules already given.

When typewriters are used it is well to avoid condensed or extra bold type faces which tend to give a heavy, gothic look. By contrast, some of the modern light and airy faces are too "open" to come across on a screen. Needless to say, the type face must be clean and free from blemishes. A plastic-carbon ribbon must be used to get a uniform deposit on the paper. Many A-V departments like special matt paper, which is expensive. In my experience this tends to flake under the hard blow of an electric machine and a well-backed heavy manuscript paper is better.

Pressure adhesive lettering tends to crack, particularly as stocks age, and must be accurately laid. Faint grey graph paper is an aid to this. For absolute certainty that the lines will not show on the background, the paper can be

reversed and illuminated from below by laying it on a horizontally placed
X-ray screen. Alternatively, the letters can be stuck to a transparent overlay
on the paper, but this does unfortunately increase their tendency to crack
off during transit to the A-V department, and storage of the originals
(say, for subsequent reproduction in another place or for amendment) is
less easy.

Mounting and labelling

A rigid mount of metal or plastic with glass faces should *always* be used.
Cardboard mounts do not protect the emulsion and become dog-eared so
risking their irreversible ingestion into the maw of an angry automatic
projector; the jam that results always involves the most important data
slide so making it difficult for even the most resourceful speaker to proceed.
Furthermore, heat expansion causes a cardboard mounted slide to "pop"
and deform. By contrast rigid mounts do not permit this to happen and
their reasonably constant dimensions make it likely, though not certain,
that, once the first is in focus, the whole sequence will be. It is still
possible to make a less than satisfactory job of mounting if care is not
taken to align the film carefully. The result is a crooked slide, perhaps with
a rim of light. A transparency mount should be chosen which has a positive
method of alignment.

Everyone is now familiar with the international labelling convention
(Fig. 16). A number in the right hand corner is an additional refinement

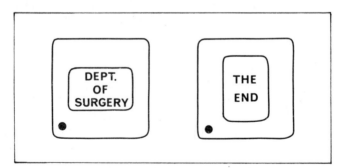

Fig. 16 International convention for labelling slides.

which can rapidly restore order when you or the projectionist drop the
whole collection. Pressure adhesive numbers are easily obtained for this
purpose.

Finally, carry slides in a rigid container or, in my view better still, in a wallet, ordered and ready to insert into the projector or its magazine. If possible, supervise the latter yourself - then you cannot blame anyone if they are out of order.

REFERENCES

1. Calnan, J. and Barabas, A. (1973) *Speaking at Medical Meetings: A Practical Guide.* London, Heineman.
2. Medawar, P. (1964) (See Chapter 2 references.)
3. Ziman, J.M. (1969) Information, communication, knowledge. *Nature,* **224**, 318.
4. Fox, T.F. (1964) *Crisis in Communication.* London, Athlone Press.
5. Pickering, G. (1961) Language, the lost tool of learning in medicine and science. *Lancet,* **2**, 115-119.
6. Dudley, H. (1969) Tasks for clinicians; the organisation and direction of clinical research. *Med.J. Aust.,* **1**, 103.
7. Simmonds, D. (1976) Medical chartist's dilemma. *Medical and Biological Illustration,* **26**, 153-158.

Appendix 1
General references and scientific writing

Some generally useful sources on writing other than those quoted in the text.

Barabas, A. and Calnan, J. (1973) *Writing Medical Papers: A Practical Guide.* London, Heinemann.
  Like their speaking guide - see below - a relatively light-hearted approach to the matter. Purists may object, but here is a wealth of practical advice.
Baron, D.N., Broughton, P.M.G., Cohen, M., Lansley, T.S., Lewis, S.M. and Shinton, N.K. (1974) The use of SI units in reporting results obtained in hospital laboratories. *Journal of Clinical Pathology,* 27, 590-597. Reprint No.3.
Booth, V. (1975) *Writing a Scientific Paper.* London, Biochemical Society, by permission of Koch Light Laboratories Ltd.
  By contrast to Barabas and Calnan this is distilled wisdom. None the worse for that and still easy to read. It is very cheap, particularly if bought in bulk and should be a "hand out" for any research worker.
British Journal of Surgery (1975) Artwork for the B.J.S. Offprint from Wright and Son, Bristol.
  By no means the last word on this subject but a help towards clarity.
Calnan, J. and Barabas, A. (1972) *Speaking at Medical Meetings: A Practical Guide.* London, Heinemann.
  See Barabas and Calnan (1973) above.
Cooper, B.M. (1964) *Writing Technical Reports.* 2nd ed. London, Pelican.
  This book was not intended for medical men. Nevertheless, it contains much that is useful. Perhaps it could help most those whose job it is to communicate through channels other than scientific speaking or writing. Committee members could read it with profit.
Council of Biological Editors (1972) *CBE Style Manual,* 3rd ed. Washington, American Institute of Biological Sciences.
  One of the most comprehensive guides to accuracy, but a little dated because of SI units - see Royal Society Symbols Committee (1971) and Baron *et al.* (1974).
Ellis, G. (1972) *Units, Symbols and Abbreviations.* London, Royal Society of Medicine.
  Currently being reworked, this is a very useful short guide.
Fowler, H.W. (1965) *A Dictionary of Modern English Usage.* Revised by

Gowers. Oxford, Oxford University Press.
This great classic is fun chiefly because it attacks the pedants. Not
that it is without this bias itself; but usually the arguments are logical.
For self-protection if nothing else, it should be on everyone's shelf.
Gowers, E. (1973) *Complete Plain Words*. 2nd ed, by Fraser, B. London,
HMSO.
See Fowler above.
O'Connor, M. and Woodford, F.P. (1975) *Writing Scientific Papers in
English*. Amsterdam, Elsevier.
This recent work makes me wonder if I should have attempted to write
anything more. However, it is a general text for scientific writers rather
than peculiar to medicine (or at least that is my rationalisation).
Contains a valuable list of abbreviations and a splendid glossary of
words and phrases to avoid. Also a very extensive reference list.
Roberts, F. (1960) *Good English for Medical Writers*. London, Heinemann.
This book, now to be found only in libraries, is a pleasant read rather
than a definitive text.
Royal Society Symbols Committee (1971) *Quantities, Units and Symbols*.
London, Royal Society.
Royal Society (1974) *General Notes on the Preparation of Scientific
Papers*. London, Royal Society.
Definitive, stately, general and a mite turgid, but useful nevertheless.
Strunk, W., Jr. and White, E.B. (1972) *The Elements of Style*. 2nd ed.
Riverside, N.J. (U.S.A.), MacMillan.
Widely quoted but perhaps less widely read; a good if slightly diffuse
book, the result of useful collaboration between disciplines.
Wilson, G. (1965) Guidance in preparing the typescript of scientific papers.
Monthly Bulletin. Pub. Health Lab. Service, **24**, 280.
See Royal Society above.
Zollinger, R.M., Pace, W.G. and Kienzle, G.J. (1961) *A Practical Guide
for Preparing Medical Talks and Papers*. New York, MacMillan.
Out of print as far as I am aware, but if your library has a copy, well
worth consulting because of its instructions on visual aids.

Appendix 2
Reference conventions

Recommendations

The following recommendations for the citation of references have been drawn up by members of the Medical Section of the Library Association to serve as a guide for editors, authors and others concerned with medical and scientific publication. They agree closely with the basic principles specified in the latest draft revision of British Standard 1629.

A bibliographical reference should not be included in a publication unless (a) it is intended as confirmation of, or is relevant to, a statement made by the writer; (b) it refers to further literature on the matter under discussion.

Definition

A bibliographical reference is a set of data describing a publication or part of a publication. The elements forming a reference must be sufficiently detailed to identify it without difficulty.

Methods of citation

Two main systems of reference citation are in common use:
1. A system in which superior or bracketed numbers are used in the text and references to them are given as footnotes or collected together at the end of the publication.
   Examples: Smith[3]
              Smith (3)
              It has been shown[3]
2. The so-called "Harvard" system, in which names and dates are given in the text with references arranged alphabetically by the authors at the end of the publication.
   Examples: Smith (1963)
              Smith and Jones (1964)

When two or more references to an author are cited they are arranged in chronological order. When several papers by the same author(s) in one year are cited, these are distinguished by the addition of a, b, etc. after the date.

86

When reference is made to a publication by three or more co-authors, all names are quoted in the first citation. In subsequent reference only the first name need be quoted, followed by *et al.,* but if such omission would cause confusion (e.g. papers by Smith, Brown and Jones (1963) and by Smith, Jones and Brown (1963) might both be quoted in the same paper) all names should be given.

Examples (first reference): Smith, Brown and Jones (1963)
           (later reference): Smith *et al.* (1963)

The Harvard system is recommended as it permits the last-minute addition or removal of references without the necessity for renumbering or rearrangement. This system is used for the examples given below.

The elements

A bibliographical reference should preferably include the following elements, in the order given if the Harvard system is used, or with date at position shown by the asterisk (*) if another system is used.

Journal article	*Book or part of a book*
author(s)	author(s), compiler(s), editor(s) or contributor(s)
date of publication	date of publication
title of paper	title of book (and, when applicable, title of contribution)
journal title, *, volume	edition number or other specification of the edition
pagination	place of publication
	publisher, *
	pagination (in a reference to part of a book)

The number of elements used and the order in which they are given will depend on the style and system of references used by the journal or book in which the publication is to appear.

Name (s) of author (s), editor (s), etc., followed by initials. "Jr.", "Sr.", and similar descriptions may normally be omitted. In references to editorial or other anonymous contributions to periodicals the name of the journal should be quoted as author. In references to anonymous books, the title is the first element. Editor (s) may be indicated by (ed.) or (eds.).

Year of publication, in parentheses. The addition of a, b, etc. is necessary if several papers by the same author(s) in one year are cited.

Title of paper. This is considered essential as providing information on the relevance of the citation. Titles in Cyrillic or Greek alphabets should be transliterated. British Standard 2979 *(Transliteration of Cyrillic and Greek characters)* includes (Section 1) transliteration of Cyrillic according to the "British system", which is used in *British Union Catalogue of Periodicals (BUCOP)* and also in *Index Medicus.* Section 3 gives a system of transliteration from Greek based on international practice. Alternatively these titles, as also those in other foreign languages, may be translated into English and printed within square brackets. In such cases it is advisable to indicate the language of the text and the existence of an English or other summary.

Book. This should conform to what is printed on the title page. For non-serial publications appearing in more than one numbered volume the abbreviations "vol." may be inserted.

Journal. Most periodicals use a recognised system of journal title abbreviation. The system adopted for *Index Medicus* since January 1971 is strongly recommended. It embodies the abbreviations in the *International List of Periodical Title Word Abbreviations* published in 1970 by the UNISIST/ICSU-AB Working Group on Bibliographic Descriptions and, with additions, by the *American National Standards Institute* in 1971 as *National Clearinghouse for Periodical Title Word Abbreviations: Word-Abbreviation List.* The forthcoming BS 4148: *The Abbreviations of Titles of Periodicals. Part 2: Word Abbreviation List* will also be substantially the same as the *International List.* It is also used for *BUCOP* . Moreover, the extensive list of journals indexed in *Index Medicus,* giving both full title and abbreviation, is continuously updated and is available separately at a modest price. The use of *ibid., idem,* etc. should be avoided.

The volume number of a journal is to be given in Arabic numerals, even if the original publication uses Roman numerals.

Pagination. Inclusive pagination is to be encouraged as it gives an indication of the length of a paper. When given, the page numbers should be written in full, e.g. 301-326, not 301-26.

Place of publication: publisher. Place of publication is an important element; it may distinguish between different printings of the same book in different countries, possibly with differences in the text and of the date. Place of publication is given as it appears on the title page. Where two or more places of publication or publisher's names appear on the title page, only the first is given in each case.

Punctuation in lists of references is not necessary between date and title or between journal title and volume number (except a full point after an abbreviated word). A full point should be placed after the last page number.

Parts of volumes of periodicals (numbers or dates) should be included in references only where pagination is not continuous throughout the volume (e.g. in the case of some Russian journals). In other cases they become superflous as soon as the journal is bound.

Examples

References to periodicals

Burke, D.C. (1961) The purification of interferon. *Biochem. J.,* **78,** 556-563.
Crick, F.H.C., and Watson, J.D. (1956) Structure of small viruses. *Nature (Lond.),* **177,** 473-476.
Mikhailov, F.A. (1970) On the diagnosis of tuberculous masadenitis in adults. In Russian. English summary. *Klin. Med. (Mosk),* **48,** No. 8, 78-81.

In an anonymous paper where the name of the journal is used for quotation, it is not repeated:

British Medical Journal (1969) Psychogenic dyspnoea. **4,** 382.

For journals without volume numbers the page numbers are indicated by p.

Islip, P.J., and White, A.C. (1964) Some reactions of 2-(3-oxindolyl) ethylamines. *J.Chem. Soc.* p.1201-1204.

References to non-periodical literature

Books:

Lane-Petter, W. (1961) *Provision of Laboratory Animals for Research; A Practical Guide.* Amsterdam, Elsevier.
Pazzini, B. (1941) *La Medicina Primitiva.* Milano, Arte e Storia.
Dubos, R.H., and Hirsch, J.G. (eds.) (1965) *Bacterial and Mycotic Infections of Man.* 4th ed. London, Pitman Medical.
Deuel, H. (1957) *The Lipids.* 3rd ed., vol. 2. New York, Interscience. p. 9-12.

Contribution to books:

Porterfield, J.S. (1964) Interference and interferon. In: Harris, R.J.C. (ed.) *Techniques in Experimental Virology.* London, Academic Press. p. 305-326.

Contributions to symposia:

Gray, J.A.B. (1962) Coding in systems of primary receptor neurons. In: Beament, J.W.L. (ed.), *Biological Receptor Mechanisms.*

Symposia of the Society for Experimental Biology, No. 16. Cambridge, University Press, p. 345-354.

Congress proceedings:

Cornforth, J.W., Cornforth, R.H., and Mathew, K.K. (1959) A new stereo-selective synthesis of olefins and its application to the synthesis of all-trans-squalene. In: Mosettig, E. (ed.), *Biochemistry of Steroids. Proceedings of the Fourth International Congress of Biochemistry, Vienna, 1-6 September 1958,* vol. 4, p. 59-60.

Apthorp, G.H., and Lehmann, H. (1963) The management of sickle-cell anaemia during pregnancy. In: *Proceedings of the IXth Congress of the European Society of Haematology, Lisbon, 1963.* Basel, Karger. p. 450.

Lofts, B. (1965) Seasonal lipid changes and their possible significance in the testis of Anura. In: *Proceedings of the Second International Congress of Endocrinology, London, 17-22 August 1964.* Amsterdam, Excerpta Medica. p. 100-105. (*Excerpta Med. Int. Congr. Ser. 83.*)

Series details should be included when they assist in identification of the reference, as in the example preceding, or that following.

Sainsbury, P. (1955) *Suicide in London.* London, Chapman and Hall. (Maudsley Monograph No. 1.)

Appendix 3
Abstracts for the Surgical Research Society

Instructions for preparing abstracts

IMPORTANT: This instruction sheet must be strictly adhered to and members submitting summaries are requested to ensure that they conform to the following requirements.

1. *Ten (10) copies* of each summary should be submitted. The text must *not* exceed 200 words in length and should be typed in *double spacing on one A4 sheet* (A reference in text = 1 word)
2. The *title* of the paper should precede the text on every copy.
3. On *three copies only* the author's (authors') initials and surname(s), in that order, should be given (for women, one forename should be set out in full), but *not* titles or degree. In joint authorship the name of the person who will read the paper must be placed first. If none of the authors is a member of the Society, the name of the member introducing the paper must be added, in brackets, after the name(s) of the author(s). Also, on the same three copies only, the department, institution and place (e.g. London) should be stated. Post codes will not be included in the published summaries.
4. Indication of grant support will not be published in abstracts and should not be included.
5. *Important footnote:* At the end of the *top* copy of each summary, the following note should be added:
    - (a) The work described in this summary has/has not been previously published. (If already published, please give name of journal.)
    - (b) The work contained in this summary has/has not been read at a scientific meeting (if so, please give details).
    - (c) The work described in this summary has already been submitted for consideration of another scientific society (please specify).
6. The summaries of papers selected for the meeting will be published *unaltered* in the *British Journal of Surgery (B.J.S.)* and a preprint of these will be available at the meeting. Thus, it is essential that:
    - (a) Data are included and results and conclusions are clearly described; phrases such as "the results of these studies will be described" are not acceptable.
    - (b) The House style of the *B.J.S.* is closely observed.

91

A recent issue of the *B.J.S.* should be consulted but the following are important:

(i) The title is typed in *lower case* and *single spacing* without the use of capitals except for the initial letter. It should *not* be underlined.

(ii) Author's (Authors') name(s) is typed in *capitals.*

(iii)References should be kept to a minimum and must not exceed four. The Harvard system is used. In the text, this means Brown and Smith, 1949, preferably in a way that can be bracketed; for more than two authors, Brown *et al.,* 1950 should be used. At the end of each summary the references are listed in alphabetical order with names and initials of all authors, year of publication in brackets, title of journal (abbreviated according to *Index Medicus*) volume number underlined, and first and last pages. In the case of books, in addition to the author(s), the year of publication, title, place of publication and publisher's name must be given. Further details of the Harvard System can be found in Reference Citation Recommendations, issued by the Library Association (Medical Section), 7 Ridgmount Street, London WC1E 7AE.

7. *Textual instructions*

(a) Items to be referred to subsequently in abbreviated form should be indicated first in full unless the abbreviation is an agreed one. A list of agreed abbreviations will be circulated from time to time. Abbreviations used to characterise animals (e.g. CBA, NZB) need not be explained in full. Abbreviations are typed without full stops, e.g. DVT not D.V.T. (but lower case abbreviations have stops, e.g. s.e.m.). Subscripts to abbreviations should not be used, e.g. PAO(Pg) is preferable to PAO_{pg}.

(b) Proper names should be used for drugs with, if more generally known, the trade name in brackets: e.g. aprotinin (Trasylol).

(c) S.I. nomenclature for quantities should be used wherever possible. Authors should refer to: *Quantities, Units and Symbols* (1971), The Royal Society, London; Guide to authors, *Biochem. J.* (1973), **131**, 1 - 20; Baron, D.N. (1974) S.I. units, *B.M.J.,* **1**, 509 - 512.

8. Members who introduce papers are responsible for ensuring that summaries conform to these requirements.

9. Submission of summaries that do not conform to these requirements may result in the rejection of papers that are otherwise suitable.

10. A summary in the appropriate form is included.

IMPORTANT Please indicate to whom notice of rejection or acceptance should be sent, in a covering letter accompanying the abstract submissions.

Example

on 3 copies only { The interpretation of plasma gastrin results - an explanation D.G. Ponsford, R.M. Nixon and Margaret Smith (H.F. Jones) Surgical Unit, St Margaret's Hospital Medical School, London } on 3 copies only

The results of radio-immunoassay of plasma gastrin levels in the basal state vary widely (Hansky and Korman, 1973). Two factors may be responsible: first, different antibodies used by different workers; secondly, the specificity of antibody for big (BG) and little (LG) gastrin. The latter will determine the absolute basal level in that BG predominates in the fasting state and LG in the stimulated phase (Yalow and Berson, 1971).

We report here the cross-reactivity of three antibodies (Gas 8, Gas 9 and RCS 8) for BG and LG, and the values for basal plasma gastrin in 10 normal, 15 duodenal ulcer and 6 postsurgical patients. We have also studied the 6 postsurgical patients after stimulation by a standard meal.

The cross-reactivity for BG was Gas 8, Gas 9, 45 per cent; RCS 8, 68 per cent. Basal plasma gastrin was similar for Gas 8 and 9 but high for RCS 8. The values for incremental response showed no differences. These results suggest that the basal plasma gastrin level is higher the greater the cross-reactivity with BG, and confirms that after a meal the rise is mainly due to LG. Plasma gastrin results cannot be interpreted without knowing the component of gastrin being measured.

| Antibody | Normal subjects (pg/1 x 10^{-3}) | Duodenal ulcer patients (pg/1 x 10^{-3}) | Meal Stimulus | |
			Gastrin response (ng)	Incremental response (ng)
Gas 8	13.7 ± 1.0	18.8 ± 2.0	8.3 ± 1.5	3.8 ± 0.7
Gas 9	13.9 ± 0.6	17.6 ± 1.4	8.1 ± 1.1	3.1 ± 0.5
RCS 8	28.1 ± 3.1	29.6 ± 2.8	12.5 ± 2.1	3.1 ± 0.6

All results: Mean ± s.e.m.

Hansky, J. and Korman, M.G. (1973) Immunoassay studies in peptic ulcer. *Clin. Gastroenterol.*, 2, 275-291.

Yalow, R.S. and Berson, S.A. (1971) Further studies on the nature of immunoreactive gastrin in human plasma. *Gastroenterology*, 60, 203-214.

The work described in this summary has not been published previously, read at a scientific meeting, nor submitted to another Society. (On top copy only)

Appendix 4
Statistical sources useful to the medical biologist

I have not listed this bibliography in alphabetical order because I think it is more useful to the reader to see how he can *develop* his reading in this difficult but essential field. One of the things to avoid is the acquisition of knowledge of *tests* without at least some understanding of their underlying logic.

Moroney, M.J. (1973) *Facts from Figures.* London, Penguin.
  The amateur's vade-mecum. Good on parametric tests and control. Less helpful on the problems which confront those in the biological sciences. Helps if one forgets - as I do - simple formulae.
Langley, R. (1968) *Practical Statistics.* London, Pan.
  This is an Australian book and very well written; it does all and, for non-parametric statistics, more than Moroney's *Facts from Figures.*
Documenta Geigy (1970) Ed. Diem, K. and Lentner, C. Basle, Geigy.
  Quite the most comprehensive compedium on simple statistical logic that I have ever come across. Each word must be read; there are no short-cuts, but if one emerges at the end feeling that one has followed the reasoning, then a grasp of statistical inference is assured.
Siegel, S. (1956) *Non-parametric Statistics for the Behavioural Sciences.* Tokyo, McGraw-Hill.
  This has long been the guide-book of psychologists. It is a lucid, if somewhat thickly written, account of the available non-parametric tests plus some useful information on scales and statistical inference. Required reading for anyone working in the biological sciences. Its popularity and utility is attested by the fact that I have lost three copies.
Rigg, D. (1963) *Mathematical Approach to Physiological Problems.* New York, Williams and Wilkins.
  I fear this book is out of print. Anyone who can obtain a copy will find it rich in mathematical analogy and a fund of common sense. The statistics are minimal, but the rigour of the approach to inference and the application of mathematical problems make it a worthwhile read.
Mather, K. (1946) *Statistical Analysis in Biology.* London, Methuen.
  This book is also out of print but it contains the only comprehensible (to the non-mathematician) explanation of the inter-relationship between the various parametric tests.
Armitage, P. (1971) *Statistical Methods in Medical Research.* Oxford, Blackwell.

Professor Armitage is in the unusual position of being a highly regarded professional statistician and having at the same time allied himself with medicine. Like Mather and to a lesser extent Siegel, it is a hard read. However, the more one peruses it, the more one realises that it is a beautifully organised text. Essential to all department members who require to understand what they are doing. His *Sequential Analysis* - now in a second edition (1976) - is also illuminating if one reads slowly.

There is then a big gap before one gets to the professional texts such as:
Fisher, R.A. (1960) *The Design of Experiments.* 7th ed. Edinburgh, Oliver and Boyd.
  Though almost devious in its logic this book has some delightful introductory material that should bolster the optimism of the amateur.
Fisher, R.A. (1963) *Statistical Methods for Research Workers.* 13th ed. Edinburgh, Oliver and Boyd.
Kendall, M.G. and Stuart, A. (1963) *The Advanced Theory of Statistics.* London, Griffin.
  Like many other texts this is incomprehensible to the average layman. The best bridge between simple and advanced that I have found is:
Gilbert, N. (1973) *Biometrical Interpretation.* Oxford, Clarendon Press.

Index